"For too long tourism has been touted as a panacea, an industry which has overwhelming benefits for development of destinations. This book is ground-breaking in its efforts to expose the serious issue of land-grabbing in tourism locales, and its impacts on the more vulnerable in society. An impressive 31 case studies from around the world are provided. Professor Neef has laid out a clear argument for why this must be challenged – whether the land grabbing is at the hands of conservation agencies, governments, or multinational hotel chains – while also constructively demonstrating what instruments can be used to protect communities from future displacement. The book is a must-read for anyone committed to just, sustainable development of tourism destinations."

—Regina Scheyvens, Professor in Development Studies, School of People, Environment and Planning, Massey University, New Zealand

"This monograph is a highly important addition to tourism studies. At a time of surging debates about social justice, Tourism, Land Grabs and Displacement exposes simplistic views of tourism's development potential. It fills an important gap in our understanding of what tourism 'does' on the ground, and asks for accountability of nation states and corporations. With case studies from all over the world, this carefully researched volume should be obligatory reading for anyone working in tourism."

—Stefan Gössling, Professor of Tourism Research, School of Business and Economics, Linnaeus University, Sweden

"This book provides a powerful and timely analysis of more than 30 case studies that show how the global tourism industry and its host governments are driving land grabbing and displacement with devastating impacts on communities in the Global South. Andreas Neef provides intriguing and critical insight into the actors, discourses, mechanisms and practices behind tourism-related land grabs and discusses resistance movements and potential remedies. In these uncertain times of a global tourism crisis – with potential for substantial restructuring of the industry – this book presents a key resource for academics, students, practitioners and policy-makers in the field of tourism and development."

—Susanne Becken, Professor of Sustainable Tourism, Griffith University, Australia

"Tourism is often portrayed by its boosters as a hugely beneficial and smokeless industry that contributes to economic and environmental wellbeing. This book provides another line of critique to the unthinking promotion of tourism by highlighting the impacts of tourism-related land and resource grabbing. In doing so it exposes the way in which tourism, even when promoted as supposedly being sustainable, can displace communities, destroy livelihoods and enclose space. The book is therefore a welcome addition to the literature that critically assesses tourism and should be read by scholars, students and policy-makers alike."

—C. Michael Hall, Professor in Marketing and Tourism, University of Canterbury, New Zealand

"As the first major book on tourism and land grabbing, and containing fascinating cases from around the world, this thought-provoking book draws much needed attention to a major issue of international concern. Providing valuable new insights into why tourism should be considered an extractive industry, this book is of high importance for students and researchers seeking to understand the 'darker' side of contemporary tourism."

—Hazel Tucker, Department of Tourism, Te Kura Pakihi – Otago Business School, University of Otago

"Finally, a book that explores the land grab debate and tourism development. From all-inclusive resort tourism to wildlife tourism, Neef exposes the reality for local land owners and communities using case studies spanning the 'Global South'. A must read for anyone interested in developing a truly sustainable and responsible tourism industry."

—Julia Jeyacheya, Senior Lecturer in International Development, Future Economies University Research Centre, Faculty of Business & Law, Manchester Metropolitan University, UK

Tourism, Land Grabs and Displacement

This book examines the global scope of tourism-related grabbing of land and other natural resources.

Tourism is often presented as a peaceful and benevolent sector that brings people from different cultural backgrounds together and contributes to employment, poverty alleviation, and global sustainable development. This book sheds light on the lesser known and much darker side of tourism as it unfolds in the Global South. While there is no doubt that tourism has been an engine of economic growth for many so-called developing countries, this has often come at the cost of widespread dispossession and displacement of Indigenous and non-indigenous communities. In many countries of the Global South, tourism development is increasingly prioritised by governments, businesses, international financial institutions and donors over the legitimate land and resource rights of local people. This book examines the actors, drivers, mechanisms, discourses and impacts of tourism-related land grabbing and displacement, drawing on more than thirty case studies from Latin America and the Caribbean, sub-Saharan Africa, South and Southeast Asia, the Middle East and the Southwest Pacific. The book provides solid grounds for an informed debate on how different actors are responsible for the adverse impacts of tourism on land rights infringements, what forms of resistance have been deployed against tourism-related land grabs and displacement, and how those who have violated local land and resource rights can be held accountable.

Tourism, Land Grabs and Displacement will be essential reading for students and scholars of land and resource grabbing, tourism studies, development studies and sustainable development more broadly, as well as policymakers and practitioners working in those fields.

Andreas Neef is Professor in Development Studies at the University of Auckland, New Zealand and co-editor of The Tourism-Disaster-Conflict Nexus (2018). With Chanrith Ngin, he edits the Routledge Studies in Global Land and Resource Grabbing book series.

Routledge Studies in Global Land and Resource Grabbing

Series Editors:
Andreas Neef, The University of Auckland, New Zealand
Chanrith Ngin, The University of Auckland, New Zealand

This series presents and discusses 'resource grabbing' research in a holistic manner by addressing how the rush for land and other natural resources (water, forests, minerals, etc.) is intertwined with agriculture, mining, tourism, energy, carbon markets, climate change, and disasters. This series welcomes contributions from a wide range of inter-disciplinary approaches and on a global basis.

Titles in this series include:

Industrial Tree Plantations and the Land Rush in China
Implications for Global Land Grabbing
Yunan Xu

Capitalism and the Commons
Just Commons in the Era of Multiple Crises
Edited by Andreas Exner, Sarah Kumnig and Stephan Hochleithner

Tourism, Land Grabs and Displacement
The Darker Side of the Feel-Good Industry
Andreas Neef

Agrarian Capitalism, War and Peace in Colombia
Beyond Dispossession
Jacobo Grajales

For more information about this series, please visit: https://www.routledge.com/sustainability/series/GLRG.

Tourism, Land Grabs and Displacement

The Darker Side of the Feel-Good Industry

Andreas Neef

LONDON AND NEW YORK

First published 2021
by Routledge
2 Park Square, Milton Park, Abingdon, Oxon OX14 4RN

and by Routledge
52 Vanderbilt Avenue, New York, NY 10017

Routledge is an imprint of the Taylor & Francis Group, an informa business

British Library Cataloguing in Publication Data
A catalogue record for this book is available from the British Library

Library of Congress Cataloging-in-Publication Data
Names: Neef, Andreas, 1965- author.
Title: Tourism, land grabs and displacement : the darker side of the
    feel-good industry / Andreas Neef.
Description: Abingdon, Oxon ; New York, NY : Routledge, 2021. | Series:
    Routledge studies in global land and resource grabbing | Includes
    bibliographical references and index.
Identifiers: LCCN 2020051010 (print) | LCCN 2020051011 (ebook) |
    ISBN 9780367356262 (hardback) | ISBN 9780429340727 (ebook)
Subjects: LCSH: Tourism–Developing countries–Economic aspects–Case
    studies. | Land use–Developing countries–Case studies. | Displaced
    persons–Developing countries–Case studies. | Tourism–Environmental
    aspects–Developing countries–Case studies.
Classification: LCC G155.D44 N42 2021 (print) | LCC G155.D44
    (ebook) | DDC 338.4/791724–dc23
LC record available at https://lccn.loc.gov/2020051010
LC ebook record available at https://lccn.loc.gov/2020051011

ISBN: 978-0-367-35626-2 (hbk)
ISBN: 978-0-367-76795-2 (pbk)
ISBN: 978-0-429-34072-7 (ebk)

Typeset in Bembo
by Taylor & Francis Books

For Katja and Sonja

Contents

x *Contents*

Figures

Tables

Boxes

Preface

The first time I came across tourism-related land grabs was in March 2013 during a qualitative field study in a Moklen Indigenous community affected by the 2004 Indian Ocean Tsunami along the Andaman coast of southern Thailand. The study was part of a research project entitled "Integrated Assessment of Post-Disaster Recovery of Small Coastal Communities in Asia-Pacific (Japan, Thailand, Vietnam, Fiji)" funded by the Japan Society for Promotion of Science (JSPS). The community had become a victim of disaster capitalism, whereby local government officials and corporate tourism developers colluded in dispossessing the formerly seafaring people of their customary land, including their beaches, mangrove areas, coconut groves and cemetery areas. A month later, I visited the site of a large-scale Chinese tourism project in the southwestern Cambodian province Koh Kong. More than 1,600 families had been dispossessed and displaced by the Cambodian government from the coast to the interior of the Botum Sakor National Park to make way for the tourism concession which occupies about 20 per cent of Cambodia's coastline. Having conducted research into customary resource rights, land grabbing and development-induced displacement in Southeast Asia since the late 2000s, I was appalled by the impunity with which governments and tourism investors took advantage of local communities' lack of formal land titles in both cases. Yet I was also impressed by the courage of residents who continued to resist these land grabs and attempts to displace them from their ancestral land. I made the decision to conduct more in-depth research into these processes that have received very little attention in the wider tourism literature.

In 2016, I started a new research project into the tourism-disaster-conflict nexus under an Early Staff Faculty Research Development Grant (FRDF) from the University of Auckland. The grant allowed me to conduct follow-up research on the impact of post-disaster tourism development on Indigenous communities in southern Thailand. I was also able to expand my research to the Southwest Pacific, where I examined the role of the tourism industry in the recovery from disastrous cyclone events in Vanuatu and Fiji. On Efate, Vanuatu's main island, I noticed how foreign investors had taken advantage of the small island developing state's legal framework that allowed them to lease customary land from local Ni-Vanuatu communities, thereby enclosing much of the island's coastal land. During the fieldwork in Thailand and Vanuatu, I was accompanied by my daughter Katja who acted as both icebreaker and informal research assistant. She also came along on field trips to Indonesia, Fiji, Tanzania and Peru and continuously challenged my assumptions and perspectives on the impacts of tourism and protected area management on customary land rights and Indigenous cultures.

I am grateful to Monsinee Attavanich, Preeda Kongpan, Maitree Jongkraichak and Arusa Panyakotkaew for their assistance with the case studies in Thailand. Sonia Ann Wasi helped with the fieldwork in Vanuatu. My good friend Siphat Touch facilitated the research on the Cambodia case study. Two doctoral students in our Development Studies programme at the University of Auckland, Inaz Ahmad and Dina Viktoria Sinlae, provided helpful insights into cases of tourism-related land grabbing and displacement in the Maldives and Eastern Indonesia respectively. Dina introduced me to her study area in Labuan Bajo and shared information on the impact of tourism development on female seaweed farmers in Rote Ndao, Indonesia. Julian Neef authorised me to use two of his photos from Cuzco, Peru. Peter Elstner provided the map with the case study locations.

In 2018, I was asked by Tourism Watch at Bread for the World in Berlin, Germany, to conduct a global study into tourism, land grabs and displacement, with a particular emphasis on the Global South. Staff at Tourism Watch – Laura Jäger, Antje Monshausen and Robert Wenzel – provided excellent guidance during the three-month work on the study and were instrumental in some of the case study choices. The study was completed in February 2019 and presented at the International Tourism Fair (*Internationale Tourismus-Börse – ITB*) in Berlin in March 2019. I am grateful to Minu Hemmati who moderated our panel discussion at the ITB and all participants who provided valuable feedback. I also want to thank Caroline Kruckow from Bread for the World who kindly organised a seminar for development practitioners, NGO representatives and government officials at the German Federal Ministry of Food and Agriculture (BMEL) and the Federal Ministry of Economic Cooperation and Development (BMZ). Participants at this seminar provided critical and constructive comments on my study that helped to further sharpen my thinking around tourism, land grabs and displacement.

I am grateful to the publisher for accepting my proposal to expand on my earlier study for Tourism Watch and develop it into a book publication by adding more case studies and updating existing ones. Three anonymous reviewers provided helpful comments on the original proposal. My special thanks go to Hannah Ferguson and John Baddeley at Routledge for their professional, supportive and patient guidance through the writing and production process of this book.

Several colleagues provided invaluable feedback on earlier drafts of chapters in this book and drew my attention to additional case studies and key sources. I am particularly indebted to Jamie Gillen, Sharlene Mollett, Rachel

Seoighe, Callie Vandewiele, Jo Kennett, Stroma Cole, Kris Baleva, Sumesh Mangalasseri and Kearrin Sims for their thoughtful comments and suggestions. I would like to thank Lucy Benge for helping with final proofreading. Finally, thank you to my family for their support and patience during the writing of this book.

Andreas Neef
Auckland

1   Introduction

Tourism in the Global Land Grab Debate

Purpose of the book

Tourism has arguably become one of the most important economic sectors globally and has been particularly hard hit by the ongoing Covid-19 pandemic which has wiped out millions of jobs in the tourism and hospitality industry around the globe. According to the World Travel and Tourism Council (WTTC), the tourism sector in 2019 accounted for 10.3 per cent of the global Gross Domestic Product (GDP) and one in ten jobs worldwide (WTTC, 2020). The World Tourism Organization (UNTWO) reported that 393 million more people travelled internationally for tourism between 2008 and 2017 (UNWTO, 2018). In 2019, international tourist arrivals reached the mark of 1.5 billion (UNWTO, 2020).

Tourism is often depicted as an activity that provides enormous benefits to host countries and local communities in the form of employment, foreign exchange, preservation of natural and cultural heritage, and intercultural exchange. The World Tourism Organization claims that tourism has the potential to contribute directly or indirectly to all 17 Sustainable Development Goals (SDGs), agreed upon by United Nations members states in 2015 (UNWTO, 2018). Tourism has been specifically included as targets in SDG 8 "Decent Work and Economic Growth", SDG 12 "Responsible Production and Consumption", and SDG 14 "Life Below Water: Sustainable Use of Oceans and Marine Resources" (UNWTO, 2015). The tourism sector has even been labelled as the world's 'peace industry' (D'Amore, 2009; WTTC, 2016), hence it also lays claim to addressing SDG 16 "Peace, Security and Strong Institutions".

Recently, the Chengdu Declaration on Tourism and the Sustainable Development Goals made the bold statement that

> tourism is a vital instrument for the achievement of the 17 SDGs and beyond as it can stimulate inclusive economic growth, create jobs, attract investment, fight poverty, enhance the livelihood of local communities, promote the empowerment of women and youth, protect cultural heritage, preserve terrestrial and marine ecosystems and biodiversity, support

the fight against climate change, and ultimately contribute to the necessary transition of societies towards greater sustainability.

(UNWTO, 2018, p. 37)

This glamorous representation of tourism omits the fact that the industry has also played a major role in the dispossession and displacement of Indigenous communities, ethnic minorities and the urban poor, entrenched resource conflicts, ecological destruction, and socio-economic inequality in many host countries, particularly in the so-called 'developing world', hereafter referred to as the Global South (e.g. Gurtner, 2016; Farmaki, 2017; Neef and Grayman, 2018). Contemporary tourism practices have been traced back to colonialism and imperialism, while tourism's controversial entanglements with class, gender, race, and even war and militarism have also been highlighted (Pritchard et al., 2007; Weaver, 2011; Kahrl, 2012; Gonzalez, 2013; Lisle, 2016).

This book examines the global scope of tourism-related grabbing of land and other natural resources and its diverse expressions and mechanisms, for instance, by enclosing territories, displacing communities and destroying livelihood opportunities. It tries to explain why tourism has often remained 'under the radar' in the global land grab debate and will argue that it is time to consider tourism as an extractive industry and to acknowledge that tourism practices can adversely affect the rights of legitimate owners and users of land and resources in a variety of ways. It aims to expose the most important drivers, actors, mechanisms and impacts of tourism-related land and resources grabbing. The book does not claim that tourism-related land grabs have been non-existent in countries of the so-called 'Global North'. In his fascinating book *The Land Was Ours: African American Beaches from Jim Crow to the Sunbelt South*, Andrew W. Kahrl (2012) provides a rich archival portrait of the racialised struggles over black-owned beaches and coastal resort ownership and how these transformed property relations, communities and ecosystems along the southern seaboard of the United States. Yet the major focus of the studies presented in this book is the tourism sector in the 'Global South', where most of the contemporary empirical studies on tourism-related land and resource grabbing have been conducted and where local communities – both Indigenous and non-indigenous – have proven to be particularly vulnerable to infringements on their customary and/or legally acknowledged land, resource and housing rights.

The objective of this book is to raise awareness among experts and practitioners in the field of tourism and land rights, including Civil Society Organisations (CSOs), international Non-Governmental Organisations (NGOs), political decision makers, national tourism services, and tourism businesses along the supply chain (from international travel advisors, national tourism offices and local travel agencies to tour operators and hoteliers). The publication shall provide solid grounds for an informed debate on how different actors are responsible for the adverse impacts of tourism on land rights infringements, how they can proactively avoid land grabs and displacements and how those who have violated local land and resource rights can be held accountable. The book also examines some of the existing international human rights frameworks as well as voluntary guidelines and corporate codes of conduct.

The following section will discuss tourism-related land grabs in the context of the global rush for land and natural resources. Then, gaps in the study of land grabbing and displacement in the wider 'tourism and development' and 'critical tourism studies' literature will be examined, followed by an exploration of the global scope and local contexts of tourism-related land grabs and displacement. The chapter will also explain the research design, discuss the case selection and provide the analytical framework. The final section will give an overview of the book's structure.

Tourism-related land and resource grabbing within the global land grab debate

Land and resource grabbing is not a recent phenomenon, but has re-emerged as an international issue of concern following the 2007/08 financial crisis which – in conjunction with a number of other crises, e.g. around food, fuel and climate – triggered a new global rush for land and other natural resources (Borras and Franco, 2010; Kugelman and Levenstein, 2011; Hall, Hirsch and Li, 2011; Pearce, 2012; Neef, 2014). The international limelight – from media, civil society and academia – has focused on large-scale transnational land acquisitions and leases for (1) agro-industrial plantations for food, feed and biofuels, (2) logging, mining and unconventional extraction of oil and gas (e.g. fracking), and (3) various forms of green grabbing, e.g. appropriation of forest areas that could be used to earn carbon credits under international climate mitigation regimes. While in many circles, land grabs are defined as large-scale land acquisitions and leases and often have a transnational dimension, the 2011 Tirana Declaration (see Box 1.1) adopted a broader perspective, denouncing all forms of land grabbing, whether international or national and whether perpetrated at the local level (e.g., by powerful local elites, within communities or among family members) or in the form of large-scale land deals (e.g., by multinational corporations, international hotel chains or state-owned enterprises) (International Land Coalition, 2011).

Box 1.1 Definition of land grabbing in the Tirana Declaration (2011)

Acquisitions or concessions that are one or more of the following:

1 in violation of human rights, particularly the equal rights of women;
2 not based on free, prior and informed consent of the affected land-users;

3   not based on a thorough assessment, or are in disregard of social, economic and environmental impacts, including the way they are gendered;

4   not based on transparent contracts that specify clear and binding commitments about activities, employment and benefits sharing; and

5   not based on effective democratic planning, independent oversight and meaningful participation.

Source: International Land Coalition, 2011

Tourism has been conspicuously absent from the global land grab debate, for a variety of reasons: first, in contrast to the agricultural, forest and mining sectors, the tourism sector tends to be regarded as a non-extractive industry with negligible adverse impacts on people and environments. Second, tourism businesses are often assumed to be smaller in scale than agro-industrial plantations or large mining operations and therefore regarded less as a threat to the land rights of local communities. Third, tourism development has been promoted by governments, donors, development practitioners, tourism scholars and even many NGOs as a pro-poor strategy, hence critical reporting on tourism-related land grabs and displacement is not encouraged and sometimes actively impeded. Fourth, international watchdogs (e.g. the Land Matrix, an independent monitoring initiative for global land deals) have often focused on the *transnational* dimension of land and resource grabbing, while the myriad domestic investors in tourism businesses that routinely infringe on the land rights of local communities receive less attention. Finally, many corporate land deals and state-led land grabs are attributed to other sectors, such as infrastructure (airports, roads, railways, etc.) or conservation (private wildlife conservancies, national parks, etc.), although the major driver behind the land grab may in fact be the tourism sector or at least expectations by governments and businesses that visitor numbers will rise as a result of the infrastructure or conservation project (cf. Chapters 7 and 9).

For the purpose of this book, the answer to the question of whether a tourism-related land transfer is in fact a land grab is not simply determined by its size or by the degree of legality under existing national laws or international legal frameworks. Small-scale land deals for a corporate tourism project can be labelled as 'land grabbing' when they involve some form of physical or economic displacement and/or deceitful or unethical behaviour on the part of the investor. Similarly, land acquisitions by a national or local government for a tourism infrastructure or sports mega-event project can amount to a land grab when they do not follow proper procedures of consultation and informed consent, bend the existing laws in favour of the project and/or do not conform with internationally binding human rights law. In the following sections and chapters, when the term 'land grab' is used, it also implies that – in addition to land – other types of natural resources may also be subject to grabbing, e.g. water, sand, trees or wildlife.

Tourism-related land grabbing and displacement in the 'tourism and development' and 'critical tourism studies' literatures

There is a rich 'tourism and development' literature that has examined the complicated relationship between powerful tourism actors hailing from countries in the Global North and emerging tourism destinations in the Global South. The literature ranges from critical yet hopeful studies on pro-poor tourism (e.g. Scheyvens, 2002; Telfer and Sharpley, 2008; Sharpley and Harrison, 2019), political economy/ecology approaches to tourism and development studies (Britton, 1982; Goessling, 2003; Bianchi, 2009, 2015; Mostafanezhad et al., 2016) to postcolonial and post-development critiques of Global South tourism (e.g. Hall and Tucker, 2004; Mowforth and Munt, 2016; Fletcher, 2017), What brings together these different strands of literature is an emphasis on uneven power relations among the various actors involved in tourism development and increased dependencies of 'peripheral' states in the Global South on wealthy 'core' economies in the Global North. Matthews (1978, p. 79) described tourism as a 'new colonial plantation economy' whereby "[m]etropolitan capitalistic countries try to dominate the foreign tourism market", while Crick (1989, p. 322) viewed tourism in the developing world as a form of "leisure imperialism" and "the hedonistic face of neocolonialism" (cited in Hall and Tucker, 2004, pp. 4–5). For Gonsalves (1993, p. 11) "modern tourism is an extension of colonialism (with all the attributes of a master-servant relationship)".

Yet land grabbing, dispossession and displacement associated with tourism development have rarely been the central focus of this literature. For instance, Hall and Tucker (2004) examine 'displacement' from a tourism, migration and diaspora perspective rather than exploring violent forms of tourism-related dispossession and displacement. Mowforth and Munt (2016) dedicate a small section of their book *Tourism and Sustainability: Development, Globalisation and New Tourism in the Third World* to 'displacement and resettlement' with a focus on disaster capitalism following the 2004 Indian Ocean Tsunami (cf. Chapter 5), evictions for tourism purposes during Myanmar's military rule (cf. Chapter 6) and the forced displacement of the Maasai in Kenya and Tanzania (cf. Chapter 7). In their book *Tourism and Development in the Developing World*, Telfer and Sharpley (2008) discuss tourism-related 'relocation' in a one-page sub-section, drawing on secondary sources from Mexico and China. Likewise, Nkyi and Hashimoto (2015) in their book chapter on 'human rights issues in tourism development' dedicate less than a page on 'displacement', referring mostly to Keefe and Wheat's (2008) examples from Tanzania, Kenya and Cuba.

One of the earliest case studies on tourism and displacement in the Global South was conducted in the Gambia by social anthropologists B.E. Harrell-Bond and D.L. Harrell-Bond (1979) who examined how foreign tourism investors were lured into the country by tax breaks and other incentives and with funding from the African Development Bank, the German *Kreditanstalt für Wiederaufbau* (Credit Bank for Reconstruction) and the World Bank, thereby physically and economically displacing local communities and industries. Yet the study provided information on the foreignisation of Gambia's tourism industry at the macro-level only, without delving into a deeper analysis of local-level land acquisitions, dispossessions and displacements. Similarly, political economist Stephen Britton (1982), who developed an enclave model for Third World tourism, pointed to the problems of foreign control of the tourism industry and leakage of foreign exchange earnings, drawing on examples from Fiji and the Cook Islands, two small island developing states in the Southwest Pacific. However, his analysis did not provide insights into how tourism-related land grabbing impacted the customary rights of local communities and to what extent it caused physical or economic displacement.

International attention to tourism-related land grabs and displacement in the 1990s and the first decade of the 21st century was primarily raised by international human rights advocacy groups, such as – recently deregistered – UK-based tourism watchdog Tourism Concern (e.g. Eriksson et al., 2009). Critical media outlets, such as *The Guardian* or *Al Jazeera* have also widely reported on land grabbing by foreign tourism investors in the Global South, particularly in post-disaster, post-conflict and 'conservation for tourism' contexts. Research into post-disaster land grabbing for tourism purposes was greatly influenced by Naomi Klein's notion of 'disaster capitalism' in her seminal book *The Shock Doctrine: The Rise of Disaster Capitalism* (Klein, 2007; cf. Neef and Grayman, 2018). Attention to the contentious linkages between tourism, conservation and displacement has been augmented by Brockington, Duffy and Igoe's (2008) book *Nature Unbound: Conservation, Capitalism and the Future of Protected Areas*. The authors draw attention to how ecotourism has been employed to justify and legitimate dispossession and displacement in the name of conservation.

In the 2010s, the 'critical tourism studies' literature has made important contributions to examining tourism-related land grabbing and displacement at local, regional and national levels (Cohen, 2011; Gonzalez, 2013; Devine, 2014). Conceptually, there have also been advances towards identifying some of the major mechanisms and drivers of land grabbing associated with tourism development in the Global South, most notably by critical scholars in the field of 'violent tourism geographies' (Devine and Ojeda, 2017; Büscher and Fletcher, 2017; Salazar, 2017). These scholars have linked tourism development in the Global South to violent practices of nation-building, place commodification, border securitisation and state territorialisation (e.g., Devine, 2017; Devine and Ojeda, 2017; for an overview, see Gibson, 2019). Most recently, the geopolitical dimensions of tourism-driven land grabbing and state territorialisation have also been highlighted both empirically and conceptually (e.g. Rowen, 2018; Gillen and Mostafanezhad, 2019; Mostafanezhad, 2020).

Global scope and local contexts for tourism–related land grabs

In the 2010s, the highest tourism growth rates were recorded for the so-called Least Developed Countries (LDCs) and nearly all of them – 46 out of 50

LDCs – depend on tourism as their primary source of foreign exchange earnings (Becken, 2014). Sixteen out of the 20 fastest growing travel and tourism economies in 2018 were countries in the Global South (WTTC, 2019). Yet, most of these countries are post-colonial states and many are also post-conflict nations where land governance systems have remained relatively weak and are often incoherent due to overlaps between legal regimes stemming from colonial times, Western-style land reforms undertaken by post-colonial governments and customary land tenure systems. Investors from the Global North as well as domestic political and economic elites can use such weaknesses and incoherencies in national legal frameworks to pursue their own economic agendas through tourism, often at the expense of indigenous communities, ethnic minorities, smallholder peasants, artisanal fisherfolks and other marginalised groups that lack voice in national development debates.

Illegitimate or unethical land transfers for tourism development often take place in countries that are marked by significant power differentials between actors and where host governments do not necessarily act to protect weaker groups within society (Price, 2015). A common argument that has been made that many developing countries suffer from insufficient law enforcement by the public sector which allows land grabbing in the tourism sector to occur (e.g., Ojo, 2013). Yet, in many cases, national legal frameworks in the Global South have been designed to promote domestic and foreign direct investment in tourism, to facilitate land transactions and even to make it easier for governments to drive local communities off their land (cf. Neef, 2016).

Few countries have provided a strong protective framework for customary rights to natural resources. In some countries, such as India and Indonesia, new land acquisition laws have been established that – on paper – provide better protection of vulnerable groups, e.g. by making social impact assessments mandatory (Price, 2018). Yet, there are exceptions where the new laws do not apply. Indonesia's Law 2/2012 on Land Acquisitions in the Public Interest, for instance, does not apply when the land needed for investment in the public interest is located in a forest area or when a local or national government institution wants to acquire land that is smaller than five hectares, which may be sufficient for a small tourism zone or a new sports stadium (Bakker and Reerink, 2015). Little is known about the real impact of how such new and seemingly progressive land acquisition acts are implemented, but there is anecdotal evidence that these laws have actually sped up the land acquisition process instead of leading to more participation and more effective protection of the land rights of local communities (cf. Chapter 9 for the case of greenfield airport development in Goa, India).

In some countries, progressive policies that adopt international human rights principles have yet to be incorporated into a comprehensive national legal framework. For example, the Sri Lankan Cabinet approved a National Involuntary Resettlement Policy in 2001 that was modelled on international best-practice safeguard guidelines but did not provide any legal instruments to implement it (Price, 2015). This opened the door to widespread evictions of local communities by the Sri Lankan military for tourism development over the past decade (cf. Chapter 6).

Another key contextual factor of tourism-related land grabbing is that the tourism sector is extremely diverse, fragmented and non-transparent. For example, in many countries targeted by the tourism industry, property rights may change hands several times after the construction of tourism premises. Also, ownership of the built infrastructure may be different from property to the land. This makes it easier for investors to undermine international human rights laws and more difficult for national and local government bodies to regulate the sector and protect local communities from land rights violations caused by corporate activities (IHRB and Tourism Concern, 2012).

Debates about tourism-related land grabs have often placed particular emphasis on how the rights of Indigenous peoples are being affected by the tourism sector. Yet rights infringements do also affect non-indigenous peoples that are even less protected under international human rights frameworks (cf. Chapter 11). This will become particularly evident in the case studies from Cambodia (Chapter 3), Philippines (Chapter 3), Costa Rica (Chapter 4) and Sri Lanka (Chapter 6), where non-indigenous communities have also seen their land and resource rights violently withdrawn or slowly eroded through tourism actors at government and corporate levels.

Finally, an important contextual factor of land grabs by the tourism industry in the Global South relates to the common lack of legal literacy among the affected communities. Many groups (1) have very limited knowledge of the official legal and judicial system, (2) have a conception of resource ownership that differs markedly from western understandings of land and resource property, (3) lack the financial, technical and political capacity to challenge land rights infringements, and (4) are often unaware of who they should report their grievances to.

Research approach, case selection and analytical framework

Research approach

For the purpose of this book I have used both primary and secondary sources to develop the major arguments. I have conducted field research on the topic of tourism and land grabbing in Thailand (four field visits), Cambodia (two field visits) and Vanuatu (one field visit) between 2013 and 2018. I am also familiar with tourism-related land grabs in Indonesia, Tanzania, Laos, Myanmar and Philippines due to either my own research on closely related topics or through supervision of PhD students at the University of Auckland and previously at Kyoto University. Yet the majority of cases discussed rely on the examination of secondary sources and draw on refereed international journals, media articles, personal and institutional blogs, government and NGO reports, corporate websites, and legal documents. Literature searches were undertaken via citation databases, such as Scopus and Google Scholar, using a range of relevant keywords. The emphasis of the search was on English-language academic literature but articles and documents in German, French and Spanish language were also considered.

Case selection

While there is no doubt that tourism-related land grabbing and displacement have occurred in many countries of the Global North, particularly in settler states like the United States, Australia and New Zealand, the focus of this book has been on case studies from the Global South. In selecting the case studies, I focused on (1) broad geographical coverage, (2) high actuality, and (3) trustworthiness of sources. An initial search generated more than 50 relatively well-documented case studies. This preliminary selection was reduced to 31 by eliminating cases that (1) were not based on thorough and independent investigation by more than one source, (2) did not contain new information after 2015, and/or (3) were somewhat redundant as they did not provide any additional insights into actors, discourse, mechanisms and impacts of tourism-related land grabbing.

Due to the principal focus on English-language sources, there is a potential bias towards regions and countries where more articles, documents and media reports were available in English. Nevertheless, the case studies cover a wide range of world regions, including North America, Central America, the Caribbean, South America, sub-Saharan Africa, the Middle East, South Asia, Southeast Asia, East Asia and the South Pacific. Figure 1.1 provides a map indicating the 26 countries where the 31 case studies have been conducted.

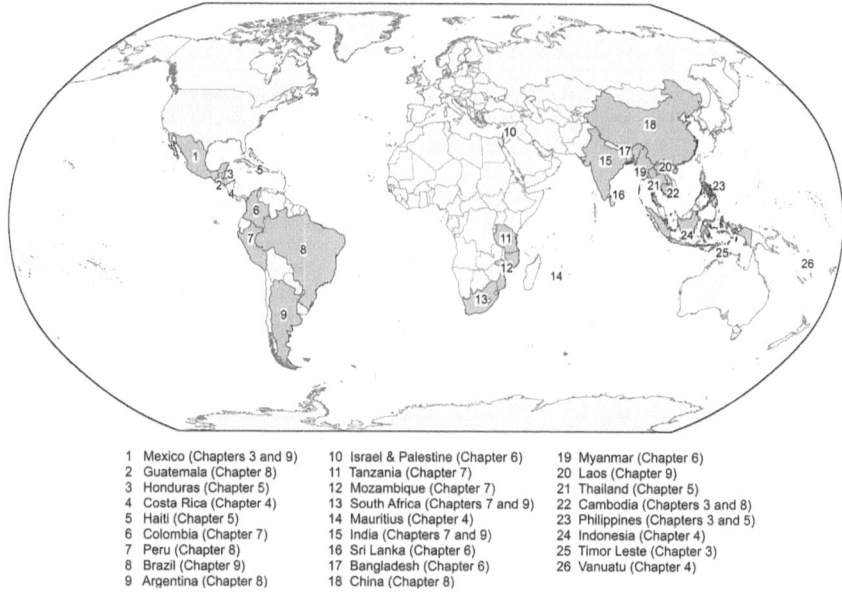

1 Mexico (Chapters 3 and 9)	10 Israel & Palestine (Chapter 6)	19 Myanmar (Chapter 6)
2 Guatemala (Chapter 8)	11 Tanzania (Chapter 7)	20 Laos (Chapter 9)
3 Honduras (Chapter 5)	12 Mozambique (Chapter 7)	21 Thailand (Chapter 5)
4 Costa Rica (Chapter 4)	13 South Africa (Chapters 7 and 9)	22 Cambodia (Chapters 3 and 8)
5 Haiti (Chapter 5)	14 Mauritius (Chapter 4)	23 Philippines (Chapters 3 and 5)
6 Colombia (Chapter 7)	15 India (Chapters 7 and 9)	24 Indonesia (Chapter 4)
7 Peru (Chapter 8)	16 Sri Lanka (Chapter 6)	25 Timor Leste (Chapter 3)
8 Brazil (Chapter 9)	17 Bangladesh (Chapter 6)	26 Vanuatu (Chapter 4)
9 Argentina (Chapter 8)	18 China (Chapter 8)	

Figure 1.1 Map with location of case studies

Analytical framework

The comparative analysis of selected case studies places particular emphasis on *practices of dispossession*, following Devine and Ojeda's (2017) conceptual ideas on violence and dispossession in tourism development. This has been subsequently developed into an analytical framework for a comparative case study by Neef et al. (2018) in southern Thailand (cf. Chapter 5). However, for the purpose of this book, the framework has been modified to include 'eviction' as an additional violent practice alongside 'enclosure', 'extraction' and 'erasure', while excluding 'commodification', 'destructive creation', and '(neo)colonialism' from the analysis. This is not to suggest, however, that these latter three practices are not at work in the cases selected for this book. The framework as outlined in Table 1.1 is examined with selected cases in Chapter 10.

Another analytical concept that is employed in this book is Harvey's (2006) notion of 'accumulation by dispossession' which he breaks down into four elements, i.e. privatisation, financialisation, management and manipulation of crises, and state redistributions. These are further explored in Chapter 10.

Table 1.1 Practices of dispossession in tourism

Type of practice	Characteristics
Eviction	Tourism physically removes communities and individuals from territories that they have previously occupied, whether under non-codified customary ownership or formally recognised communal or private land title. Eviction can occur via openly violent measures (such as burning of houses) or by more subtle means of coercion and may or may not include compensation.
Enclosure	Tourism dispossesses people from access to material means of subsistence, such as land, water, timber, fisheries and other resources. It is linked to 'accumulation by dispossession' as the tourism sector physically appropriates various types of natural resources that were previously vital to local people's livelihoods, e.g. for subsistence farming or artisanal fishery.
Extraction	Tourism development functions as an 'extractive industry' instead of being an alternative to (other) extractive industries, such as mining. The tourism sector exploits the natural environment by such practices as extracting large amounts of freshwater, removing protective mangrove forests and mining sand for beach development.
Erasure	Tourism's representational practices render pre-existing definitions of place, livelihood, identity and history invisible or erase them deliberately. The tourism sector might infringe on culturally important places (e.g. ceremonial grounds, graveyards), destroy artefacts of cultural and historic significance or render other cultures invisible through a variety of measures.

Source: Partially adapted and expanded from Devine and Ojeda (2017) and Neef et al. (2018).

Structure of the book

This book is structured as follows. In the next chapter, an overview is provided of the major actors, drivers, mechanisms, discourses and impacts of tourism-related land grabs. Chapter 3 then introduces the state as a major player in the tourism sector, given its role in delineating tourism enterprise zones, designating special economic zones, providing enabling legislation, or giving out economic land concession. Chapter 4 demonstrates how resort tourism in 'exotic' destinations has pushed host countries beyond their ecological carrying capacity and triggered a 'foreignisation of space' (Zoomers, 2010) that is often being exacerbated by residential tourism. Chapter 5 examines cases of 'disaster capitalism' (Klein, 2007), where corporations and governments have forged alliances to turn 'crisis into opportunity' and dispossessed disaster-affected communities, e.g. following the 2004 Indian Ocean Earthquake and Tsunami and the 2013 super-typhoon Haiyan in the Philippines. Chapter 6 takes a critical look at the strategies that armed forces and occupying powers have played in using tourism as a tool of displacement and cultural domination. Chapter 7 explores the interface between wildlife tourism and conservation of vast tracts of land in the form of protected areas and their impacts on dispossession and eviction. Chapter 8 looks into displacement and gentrification processes associated with (eco-)cultural heritage tourism. Chapter 9 examines how large-scale tourism infrastructure projects for sport mega-events, airports and railways have been used to legitimise compulsory land acquisitions, evictions and involuntary resettlements. Chapter 10 synthesises the principal mechanisms, practices and impacts of tourism-related land grabs and discusses some of the resistance strategies employed by affected communities. Chapter 11 assesses the potential of international legal frameworks and voluntary guidelines to better protect legitimate landowners and vulnerable communities from tourism-induced dispossession and displacement. Finally, Chapter 12 provides a brief summary of the analysis and major findings and formulates a set of policy implications and recommendations for various stakeholder groups.

References

Bakker, L. and Reerink, G. (2015) 'Indonesia's land acquisition law: Towards effective prevention of land grabbing?' in Carter, C. and Harding, A. (eds) *Land Grabs in Asia: What Role for the Law?* Routledge: London, New York, pp. 83–99.

Becken, S. (2014) 'Water equity – Contrasting tourism water use with that of the local community', *Water Resources and Industry*, 7–8: 9–22.

Bianchi, R.V. (2009) 'The 'critical turn' in tourism studies: A radical critique', *Tourism Geographies*, 11 (4): 484–504.

Bianchi, R.V. (2015) 'Towards a new political economy of global tourism revisited' in Sharpley, R. and Telfer, D.J. (eds) *Tourism and Development: Concepts and Issues* (2nd edition). Channel View Publications: Bristol, Buffalo, Toronto, pp. 287–331.

Borras, S.M.J. and Franco, J.C. (2010) 'Contemporary discourses and contestations around pro-poor land policies and land governance', *Journal of Agrarian Change*, 10 (1): 1–32.

Britton, S.G. (1982) 'A political economy of tourism in the Third World', *Annals of Tourism Research*, 9: 331–358.

Brockington, D., Duffy, R. and Igoe, J. (2008) *Nature Unbound: Conservation, Capitalism and the Future of Protected Areas*. Earthscan: London.

Büscher, B. and Fletcher, R. (2017) 'Destructive creation: Capital accumulation and the structural violence of tourism', *Journal of Sustainable Tourism*, 25 (5): 651–667.

Cohen, E. (2011) 'Tourism and land grab in the aftermath of the Indian Ocean Tsunami', *Scandinavian Journal of Hospitality and Tourism*, 11 (3): 224–236.

Crick, M. (1989) 'Representations of international tourism in the social sciences: Sun, sex, sights, savings, and servility', *Annual Review of Anthropology*, 18: 307–344.

D'Amore, L. (2009) 'Peace through tourism: the birthing of a new socio-economic order', *Journal of Business Ethics*, 89 (4): 559–568.

Devine, J. (2014) 'Counterinsurgency ecotourism in Guatemala's Maya Biosphere Reserve', *Environment and Planning D: Society and Space*, 32: 984–1001.

Devine, J. (2017) 'Colonizing space and commodifying place: tourism's violent geographies', *Journal of Sustainable Tourism*, 25 (5): 634–650.

Devine, J. and Ojeda, D. (2017) 'Violence and dispossession in tourism development: A critical geographical approach', *Journal of Sustainable Tourism*, 25 (5): 605–617.

Eriksson, J., Noble, R., Pattullo, P. and Barnett, T. (2009) *Putting Tourism to Rights: A Challenge to Human Rights Abuses in the Tourism Industry*. Tourism Concern: London.

Farmaki, A. (2017) 'The tourism and peace nexus', *Tourism Management*, 59: 528–540.

Fletcher, R. (2019) 'Ecotourism after nature: Anthropocene tourism as a new capitalist "fix"', *Journal of Sustainable Tourism* 27 (4): 522–535.

Gibson, C. (2019) 'Critical tourism studies: New directions for volatile times', *Tourism Geographies*. doi:10.1080/14616688.2019.1647453.

Gillen, J. and Mostafanezhad, M. (2019) 'Geopolitical encounters in tourism: A conceptual approach', *Annals of Tourism Research* 75: 70–78.

Gössling, S. (ed.) (2003) *Tourism and Development in Tropical Islands*. Political Ecology Perspectives. Edward Elgar: Cheltenham, Northampton, MA.

Gonsalves, P. (1993) 'Divergent views: convergent paths. Towards a Third World critique of tourism', *Contours*, 6 (3–4): 8–14.

Gonzalez, V.V. (2013) *Securing Paradise: Tourism and Militarism in Hawai'i and the Philippines*. Duke University Press: Durham, London.

Gurtner, Y. (2016) 'Returning to paradise: Investigating issues of tourism crisis and disaster recovery on the island of Bali', *Journal of Hospitality and Tourism Management*, 28: 11–19.

Hall, C.M. and Tucker, H. (2004) 'Tourism and postcolonialism: An introduction' in *Tourism and Postcolonialism: Contested Discourses, Identities and Representations*. Routledge: London, New York, pp. 1–24.

Hall, C.M. and Tucker, H. (eds) (2004) *Tourism and Postcolonialism: Contested Discourses, Identities and Representations*. Routledge: London, New York.

Hall, D., Hirsch, P. and Li, T.M. (2011) *Powers of Exclusion: Land Dilemmas in Southeast Asia*. National University of Singapore Press: Singapore.

Harrell-Bond, B.E. and Harrell-Bond, D.L. (1979) 'Tourism in the Gambia', *Review of African Political Economy*, 14: 78–90.

Harvey, D. (2006) 'Neo-liberalism as creative destruction', *Geografiska Annaler*, 88B (2): 145–158.

Higgins-Desbiolles, F., Carnicelli, S., Krolikowski, C., Wijesinghe, G. and Boluk, K. (2019) 'Degrowing tourism: rethinking tourism', *Journal of Sustainable Tourism*, 27 (12): 1926–1944.

IHRB and Tourism Concern (2012) 'Frameworks for change: The tourism industry and human rights.' Available at: www.ihrb.org/pdf/2012-05-29-Frameworks-for-Change-Tourism-and-Human-Rights-Meeting-Report.pdf (accessed 19 January 2019).

International Land Coalition (2011) 'Tirana Declaration'. Available at: www.landcoalition.org/sites/default/files/documents/resources/tiranadeclaration.pdf (accessed 10 November 2018).

Kahrl, A.W. (2012) *The Land Was Ours: African-American Beaches from Jim Crow to the Sunbelt South*. Harvard University Press: Cambridge, MA, London.

Klein, N. (2007) *The Shock Doctrine: The Rise of Disaster Capitalism*. Penguin Group: London.

Kugelman, M. and Levenstein, S.L. (eds) (2012) *The Global Farms Race: Land Grabs, Agricultural Investment, and the Scramble for Food Security*. Island Press: Washington, Covelo, London.

Lisle, D. (2016) *Holidays in the Danger Zone: Entanglements of War and Tourism*. University of Minnesota Press: Minneapolis.

Matthews, H.G. (1978) *International Tourism: A Political and Social Analysis*. Schenkman Publishing Company: Cambridge.

Mostafanezhad, M. (2020) 'Tourism frontiers: Primitive accumulation, and the "Free Gifts" of (Human) nature in the South China Sea and Myanmar'. *Transactions of the Institute of British Geographers*, 45, 434–447.

Mostafanezhad, M., Norum, R., Shelton, E.J. and Thompson-Carr, A. (2016) *Political Ecology of Tourism: Community, Power and the Environment*. Routledge: London, New York.

Mowforth, M. and Munt, I. (2016) *Tourism and Sustainability: Development, Globalisation and New Tourism in the Third World* (4th edition). Routledge: London, New York.

Mowforth, M., Charlton, C. and Munt, I. (2008) *Tourism and Responsibility: Perspectives from Latin America and the Caribbean*. Routledge: London, New York.

Neef, A. (2014) 'Law and development implications of transnational land acquisitions: Introduction', *Law and Development Review*, 7 (2): 187–205.

Neef, A. (2016) *Land Rights Matter! Anchors to Reduce Land Grabbing, Dispossession and Displacement: A Comparative Study of Land Rights Systems in Southeast Asia and the Potential of National and International Legal Frameworks and Guidelines*. Bread for the World: Berlin.

Neef, A. and Grayman, J.H. (eds) (2018) *The Tourism-Disaster-Conflict Nexus*. Emerald Publishing: Bingley.

Neef, A., Attavanich, M., Kongpan, P. and Jongkraichak, M. (2018) 'Tsunami, tourism and threats to local livelihoods: The case of indigenous sea nomads in southern Thailand' in Neef, A. and Grayman, J.H. (eds) *The Tourism-Disaster-Conflict Nexus*. Emerald Publishing: Bingley, pp. 141–164.

Nkyi, E. and Hashimoto, A. (2015) 'Human rights issues in tourism development' in Sharpley, R. and Telfer, D.J. (eds) *Tourism and Development: Concepts and Issues* (2nd edition). Channel View Publications: Bristol, Buffalo, Toronto, pp. 378–399.

Ojo, D. (2013) *Land Grabbing in Tourism: An Analysis of Drivers, Impacts and Shaping Factors of Tourism-Induced Adverse Land Deals*. Master's thesis, Hochschule für nachhaltige Entwicklung Eberswalde.

Pearce, F. (2012) *The Land Grabbers: The New Fight over who Owns the Earth*. Bacon Press: Boston, MA.

Price, S. (2015) 'Is there a global safeguard for development displacement?' in Satiroglu, I. and Choi, N. (eds) *Development Induced Displacement and Resettlement: New Perspectives on Persisting Problems*. Routledge: London, New York, pp. 127–141.

Price, S. (2018) 'Legislative paradigm shifts for involuntary people movement: An update' in *Paradigm_Shift: People Movement.* Australian National University College of Asia and the Pacific, Canberra, pp. 26–31.

Pritchard, A., Morgan, N., Ateljevic, I. and Harris, C. (2007) *Tourism and Gender: Embodiment, Sensuality and Experience.* CABI Publications: Oxfordshire, Cambridge, MA.

Rowen, I. (2018) 'Tourism as territorial strategy in the South China Sea' in Spangler, J. (ed.) *Enterprises, Localities, People, and Policy in the South China Sea.* Springer: New York, pp. 61–74.

Salazar, N.B. (2017) 'The unbearable lightness of tourism … as violence: An afterword', *Journal of Sustainable Tourism,* 25 (5): 703–709.

Scheyvens, R. (2002) *Tourism for Development: Empowering Communities.* Pearson Education Limited: London.

Sharpley, R. and Harrison, D. (2019) *A Research Agenda for Tourism and Development.* Edward Elgar: Cheltenham.

Sharpley, R. and Telfer, D.J. (2015) *Tourism and Development: Concepts and Issues* (2nd edition). Channel View Publications: Bristol, Buffalo, Toronto.

Telfer, D.J. and Sharpley, R. (2008) *Tourism and Development in the Developing World.* Routledge: London, New York.

UNWTO (2015) *Tourism and the Sustainable Development Goals.* World Tourism Organization: Madrid. Available at: http://cf.cdn.unwto.org/sites/all/files/pdf/sustainable_development_goals_brochure.pdf (accessed 2 February 2019).

UNWTO (2018) *2017 Annual Report.* World Tourism Organization: Madrid.

UNWTO (2020) *World Tourism Barometer,* 8, January. Available at: https://webunwto.s3.eu-west-1.amazonaws.com/s3fs-public/2020-01/UNWTO_Barom20_01_January_excerpt_0.pdf (accessed 26 August 2020).

Weaver, A. (2011) 'Tourism and the military: Pleasure and the war economy', *Annals of Tourism Research* 38 (2): 672–689.

WTTC (2016) *Tourism as a Driver of Peace.* World Travel & Tourism Council: London.

WTTC (2019) *Travel and Tourism: Global Economic Impact & Trends 2019.* World Travel & Tourism Council: London.

WTTC (2020) *Travel and Tourism: Global Economic Impact & Trends 2020.* World Travel & Tourism Council: London.

Zoomers, A. (2010) 'Globalisation and the foreignisation of space: seven processes driving the current global land grab', *The Journal of Peasant Studies,* 37 (2): 429–447.

2 Tourism-related land grabs

Actors, drivers and discourses

The tourism industry is not easy to define, as it is a large, complex and multi-dimensional assemblage of actors many of which would not even see themselves as being part of the industry (Mason, 2016). Major actors that may be directly or indirectly involved in tourism-related land grabs and displacement are host governments with their tourism ministries, agencies and accompanying security apparatuses, transnational tourism corporations, particularly large hotel chains, international tour operators, international financial institutions, commercial banks, sovereign wealth funds, private equity firms and international conservation organisations. Their motives can range from generation of economic growth, job creation, rural development and corporate profits to political stability, crisis recovery, resource conservation and the replacement of livelihoods that are deemed backward, illicit or environmentally destructive. Appropriation and enclosure of land and other natural resources at the expense of Indigenous communities, ethnic minority groups and other local communities are justified through the deployment of various discursive strategies.

The following section presents major actors related to the tourism sector and their motivation to get involved in the appropriation of land, enclosure of areas of touristic interest and the physical and economic displacement of rightful landowners. The second section examines the various discourses that are deployed by diverse sets of actors to legitimise the appropriation of land and other resources for tourism purposes.

Actors and drivers

In a very broad sense, tourism development and associated land grabs, dispossession and displacement in the Global South are driven by a combination of 'excess capital' and 'excessive mobilities' (Gibson, 2019). To this combination, we may also add excessive corporate power and excessive use of state force in its various forms. Transnational and domestic corporate interests merge with state power and international development actors, such as multilateral development banks, to "develop novel means to secure monopolies, contain tourists and capture profits" (Gibson, 2019, p. 4). Referring to Harvey's (1989) concept of time-space compression, Mowforth and Munt (2016) have explained how the logic of capital accumulation and circulation drives the rapid spread and expansion of tourism in the Global South.

Operating at different levels – from international to national and local – drivers of tourism-related land grabs are complex and multi-dimensional and involve a myriad of actors. Tourism is a central strategy of economic growth pursued by many *governments in the Global South* which prioritise attracting foreign capital for tourism development. Yet, as Hall et al. (2015) remind us, the question of whether nation states in the Global South should be considered as 'host countries' of foreign investors – i.e. as direct sponsors of land grabs through joint tourism ventures or indirect sponsors through the provision of key infrastructure and security – or whether they are more appropriately regarded as 'target countries' that are too weak to defend themselves against foreign tourism investors is not always easy to answer. States are not homogenous entities but composed of various institutions, agencies and factions that may have competing interests and different levels of influence on decision-making processes. However, by and large, most developing country governments tend to embrace and actively seek foreign capital flows into the tourism sector. Some governments have equipped their tourism ministries or agencies with far-reaching powers, including the mandate to expropriate rightful land-owners, as exemplified by Mexico's National Fund for Tourism Development (FONATUR) discussed in Chapters 3 and 9.

For countries with limited resources to be exploited by traditional extractive industries – as is the case with many small island developing states (SIDS) or countries with extensive coastlines – sun, sea, and sand tourism seems the obvious choice for attracting foreign direct investment (FDI) and gaining foreign currency revenues. Countries with abundant wildlife, e.g. those located in the savannah regions of eastern and southern Africa or forest-rich countries in South and Southeast Asia, aim at attracting safari tourists, adventure travellers and trophy hunters to generate foreign exchange and tax income (cf. Chapter 7). For many developing country governments tourism is a key strategy to achieve stability and consolidate state power, for example after civil conflicts, but oftentimes this comes at the expense of the land rights of former enemies and defeated communities (cf. Chapter 6). For others, tourism development is chosen as a measure to rehabilitate and rebuild a country after a major disaster, such as an earthquake, tsunami or cyclone (cf. Chapter 5). The disaster event often generates a 'blank slate' or 'tabula rasa' where new opportunities arise for national and local governments to re-appropriate land and other natural resources and allocate them to various actors within the *tourism industry*, often in opaque transactions that lack due diligence and accountability. Some governments are also motivated by the prospects of replacing subsistence livelihoods of pastoralists, fisherfolks or hunter-gatherer communities that are deemed 'primitive', 'backward' and 'environmentally destructive' by more 'modern' and ostensibly more remunerative forms of employment. Moreover, governments expect to gain better international reputation by presenting themselves as a country that is open for *cosmopolitan tourists*. These tourists tend to look for a more 'imaginary' tourism experience in an 'all-inclusive' safe enclave, with pristine beaches during the day and exotic dance performances in the evening, rather than sharing their space with 'noisy' local communities and 'unsightly' Indigenous people.

The tourism industry itself comprises a bewildering range of large, often transnational actors, medium-sized hotels, carriers and tour operators, and smaller, family- or community run tourism businesses (cf. Table 2.1). Among these actors, *transnational tourism corporations* (TTCs) are considered in the literature as having the most damaging impact on customary land and resource rights in tourism destination countries of the Global South, as they are more concerned with generating short-term profit than building long-term relationships with communities in particular destinations (Swarbrooke, 1999; Mason, 2016). Hong (1985), quoted in Mowforth and Munt (2016, p. 192), describes five major ways in which TTCs participate in the hotel industry of countries in the Global South: (1) ownership or equity investment, (2) management contracts, (3) hotel-leasing agreements, (4) franchise agreements, and (5) technical service agreements, such as market surveys and feasibility studies. Among these, management contracts are by far the most common form, accounting for nearly three quarters of TTC involvement in the Global South's hotel industry (Mowforth and Munt, 2016). Hence, TTCs, i.e. *transnational hotel chains*, such as Marriott, Hyatt and Shangri-La, are rarely directly involved in processes of land grabbing and displacement. This is left to the state apparatus in the host country which can deploy a range of legal and

Table 2.1 Tourism industry actors

	Hotels	*Carriers*	*Operators*
Large	Transnational hotel groups, e.g. Hyatt, Hilton, Six Senses, Shangri-La, Marriott, Mélia Hotels, Accor, InterContinental Hotels, Four Seasons	Internationally operating airlines, e.g. Emirates, American Airlines, Lufthansa, All Nippon Airways, Air France-KLM; Internationally operating cruise lines, e.g. Royal Caribbean International	Touristik Union International (TUI), BA Holidays, Club Med, Thomas Cook, China Travel Service
Medium-sized	National hotel groups, e.g. Malima Hospitality Services (Sri Lanka), Tata Group (India)	Regional and national carriers, e.g. Disney Cruise Line (US); Helitours (Sri Lanka), Nok Air (Thailand)	Specialist national & regional operators, e.g. Greaves Tours, Thomson Safaris
Small	Small, family-owned hotels, guesthouses, Airbnb, eco-lodges, pensions	Small carriers, e.g., car rentals, taxis, minivans, buses, ferry services, boat and helicopter companies	Service providers, e.g. tour guides, dive companies, porters, caterers

Source: Partially adapted from Mowforth and Munt (2016).

regulatory measures as well as brute force, threat of violence or withdrawal of services and livelihood opportunities to remove communities from their land and homes. Hence, transnational hotel chains may simply be free-riders of state-driven dispossession and displacement.

Transnational hotel chains are constantly looking for new, attractive locations to build large-scale tourism enclaves, particularly in places that have become enlisted as *UNESCO* World Heritage Sites (cf. Chapter 8) or those featured as "the best places to visit" on the *Lonely Planet* website. While some of the major chains have adopted a human rights policy and self-regulatory codes of conduct (cf. Chapter 11), these tend to be rather generic, focus predominantly on the operational part of the hotel business and do not make explicit reference to land rights. *International tour operators* may also bear a share of the responsibility for infringement on local people's land resource rights when they turn a blind eye to cases of land grabs and displacement by hotels and resorts or to the more systemic practices of militarisation and securitisation of holiday zones in post-conflict regions.

Medium-sized national hotel groups, such as Sri Lanka's Malima Hospitality Services, have been directly implicated in land grabs and forced displacement. While *small tourism industry actors*, such as family-owned hotels, guesthouses and eco-lodges tend to be less involved in violent land grabs and forced displacements, their combined presence in major touristic areas in the Global South can fuel the rise of local land prices and living costs that may economically displace local populations.

Tourism has become a favourite development approach for *international financial institutions* (IFIs), including the World Bank, the Asian Development Bank (ADB), the African Development Bank, the Inter-American Development Bank, the European Bank for Reconstruction and Development and the International Monetary Fund (IMF) and a major part of their lending portfolio has been dedicated to the tourism sector, including supporting infrastructure projects (Table 2.2). The Inter-American Development Bank, for instance, gave a loan of US$800 million to the Brazilian government in 2003 to create its Ministry of Tourism and to improve tourism infrastructure in the country's northeast (Telfer and Sharpley, 2008).

IFIs also play a role in providing technical assistance and guidance to large-scale tourism infrastructure developers and legitimising displacements by weighing the 'costs' of resettling a relatively small number of people against the alleged 'benefits' of the tourism sector in terms of providing job opportunities, stimulating economic growth and alleviating poverty (cf. Chapter 9 for the role of the ADB in justifying resettlements of ethnic minorities for Luang Prabang's airport expansion in Laos). Finally, IFIs may inadvertently or deliberately lay the ground for tourism investment through their support of private land titling programmes that may undermine communally held rights to land, as evidenced in the case of the Caribbean coast of Honduras (Chapter 5).

In their reports, IFIs often present successful tourism-driven growth in developing countries as showcases of what is possible when the local real estate

Table 2.2 Actors in the financialisation of the tourism sector in the Global South

International – Regional	International Financial Institutions; e.g. World Bank, International Monetary Fund (IMF), Central American Bank for Economic Integration; Asian Development Bank, Inter-American Development Bank, African Development Bank, Asian Infrastructure Investment Bank; UN organisations, e.g. United Nations Development Programme (UNDP), Commission for Africa (UNECA)
National – Government	Federal banks, e.g. Banco de México; national development banks, e.g. Export-Import Bank of China; Foundation for Investment and Development of Exports, Honduras; Overseas Private Investment Corporation (OPIC), USA; bilateral donor agencies, e.g. UK's Department for International Development (DFID), Germany's Kreditanstalt für Wiederaufbau KfW (Credit Bank for Reconstruction); Sovereign wealth funds, e.g. Temasek (Singapore), Qatar Investment Authority, Public Investment Fund (Saudi Arabia)
(Trans)national – Commercial	Internationally operating commercial banks, e.g. HSBC, Goldman Sachs, JP Morgan, Citibank, Deutsche Bank, Barclays; real estate investment trusts, e.g. Blackstone-Embassy (India); private equity firms, e.g. Blackstone Group (USA); insurance companies, e.g. Anbang Insurance Group (China); national pension funds

sector is opened up to foreign tourism investment (Zoomers, 2010). For instance, the Cape Verde Islands off the West African coastline and Mauritius in the southern Indian Ocean (cf. Chapter 4) have rapidly risen on the Human Development Index from "low" to "medium" and "high" respectively. This motivates other countries to follow these 'role models', while overlooking the dark side of their success stories, i.e. massive outmigration of Cape Verde's original population and the expansion of squatter settlements in Mauritius due to excessive land prices that have become unaffordable for local residents.

In recent years, several *government-owned investment companies and sovereign wealth funds*, such as Singapore's Temasek Holdings, Saudi Arabia's Public Investment Fund and the Qatar Investment Authority have become major players in the financialisation of the global tourism sector (e.g. Capapé, 2016). *Real estate investment trusts, private equity firms* (e.g. Blackstone Group, USA), *insurance companies* (e.g. Anbang Insurance Group, China) and *national pension funds* also have acquired major stakes in TTCs, thereby becoming indirectly involved in land grabs, enclosures and forced evictions (Table 2.2).

Aside from the major actors in the tourism industry and its financial partners, there are various *tourism interest groups* at international, national and local level (see Table 2.3). At the international level, the most important industry group that lobbies on behalf of the global tourism sector and its expansion to countries in the Global South is the *World Travel and Tourism Council* (WTTC) which receives support from some of the principal tourism companies (Mason,

Table 2.3 Tourism interest groups

Scale	Industry groups	Non-industry groups	Single interest groups
International	UN World Tourism Organization (UNWTO), World Travel and Tourism Council (WTTC), Pacific Asia Travel Alliance (PATA), World Indigenous Tourism Association (WINTA)	Environmental, social & cultural organisations, such as WWF, African Wildlife Fund, Rainforest Action Network, Conservation International, Rainforest Alliance, Survival International	Asian Peasant Coalition, Vía Campesina, Food First Information and Action Network (FIAN), Burma Action Group, Sri Lanka Campaign for Peace & Justice, Land Watch Asia, World Heritage Watch
National	National tourism organisations, e.g. Mexico's National Fund for Tourism Development (FONATUR); Chamber of Tourism (Costa Rica), Department of Tourism (Vanuatu)	Tourism Concern (UK), GOOD Travel (New Zealand), Tourism Watch (Germany), Lonely Planet (Australia), TripAdvisor (USA), Fair Trade in Tourism (South Africa)	National human rights advocacy groups, e.g. Bharat Mukti Morcha (India), Chumchonthai Foundation (Thailand), Emek Shaveh and Ir-Amim (Israel), La'o Hamutuk (Timor Leste)
Local	Regional tourism business associations, e.g. Bali Tourism Board (Indonesia); local area promotion partnerships, e.g. the Mirador Basin Project (Maya Biosphere Reserve, Guatemala)	Local government bodies, e.g. Authority for the Protection and Management of Angkor – APSARA (Siem Reap, Cambodia); residents' associations, e.g. La Voz de Guanacaste (Pacific Coast, Costa Rica)	Local groups opposed to tourism development, e.g. Citizen Action for Île à Vâche (Haiti), Kilusang Magbubukid Ng Pilipinas (Philippines), Manggarai Student Alliance (Indonesia), Goa Foundation (India)

Source: Adapted from Mowforth and Munt (2016).

2016). Its directors feature former government officials in the tourism sector; the current WTTC president Gloria Guevara, for instance, is a former Secretary of Tourism for Mexico. Similarly, the *UN World Tourism Organization* (UNWTO) has become an influential promoter of tourism development in the 'developing world'. Its current executive director is Zhu Shanzhong, a former vice-chairman of China National Tourism Administration (CNTA). These international lobby groups have turned a blind eye to tourism-related land grabs and evictions and have contributed their fair share to some of the discourses that have been deployed to legitimise them.

Other *international non-industry groups*, such as Pacific Asia Travel Alliance (PATA) and World Indigenous Tourism Association (WINTA), have tried to raise awareness about the global tourism sector's threat to Indigenous land rights. PATA and WINTA (2015) have been the driving forces behind the 2012 Larrakia Declaration with its six principles calling for the tourism sector's respect for customary land and water, law and traditional knowledge, traditional cultural expressions, and cultural heritage as well the protection and promotion of Indigenous culture through well-managed tourism practices and appropriate interpretations (cf. Chapter 11).

International conservation organisations, such as the African Wildlife Foundation, Conservation International and the World Wide Fund for Nature (WWF), have been criticised for siding with national governments to drive conservation- and tourism-related resettlements and for providing the discursive legitimation of these practices by denouncing original dwellers in protected areas as 'eco-threats' (cf. the case of Hacienda Looc in the Philippines in Chapter 3, the dispossession of Garifuna communities on the North Coast of Honduras in Chapter 5 and the eviction of the Maasai from the Serengeti National Park in Chapter 7). Yet other conservation organisations, such as the Rainforest Alliance or the Rainforest Action Network act as critics of state-driven and corporate tourism development that comes at the expense of community-based tourism initiatives and the customary rights of Indigenous people (e.g. the case of the Maya Biosphere Reserve in Guatemala in Chapter 8).

Among non-industry actors at the national level there are a number of *tourism watchdogs and pressure groups*. One of the most influential groups in terms of raising awareness about tourism-related land grabs, displacement and other human rights abuses was the UK-based organisation Tourism Concern until its closure in September 2018. Other important groups in this category are Tourism Watch in Germany and South African pressure group Fair Trade in Tourism. *Single interest groups* at various geographical levels play a major role in garnering resistance, providing advocacy and organising protest movements. Among these are transnational social movements, national human rights advocacy groups and local groups resisting neoliberal tourism development and associated land grabs, dispossession and displacement.

In recent years concerns have been raised among civil society organisations and targeted communities regarding actors that are usually not directly associated with the tourism industry. These include *armed forces* (military, paramilitary units and police) and *private security firms* that in some countries have played an increasingly prominent role in the securitisation of the tourism sector as a whole or for a particular resort-complex or tourist zone. More disturbingly, military forces have also taken a direct commercial interest in the tourism industry and have been involved in the violent dispossession and eviction of Indigenous communities and ethnic minorities and sometimes former enemies (see the cases of Myanmar (Burma), Sri Lanka and Bangladesh in Chapter 6).

Discourses

Tourism development and associated land grabbing and displacement are justified through a range of discursive strategies. In the following, the most prominent discourses are discussed, namely the (1) public purpose discourse, (2) idle land discourse, (3) tourism for conservation discourse, (4) tourism and security discourse, and (5) crisis discourse.

Governments that prioritise the tourism sector on the grounds of economic growth and poverty alleviation have often adopted a *'public purpose' discourse* to justify small- and large-scale evictions and resettlements of communities from areas with high tourism potential (cf. Chapter 3 for the case of the Philippines). While 'public purpose' sounds innocuous and well-meaning, it tends to become a misrepresented term in many official government discourses. Once tourism development is accepted by wider society as a 'public purpose', governments can then invoke the concept of 'eminent domain' as a legal mechanism to take away private or communal property from rightful owners. In most cases, the benefits from such forced land acquisitions for 'public purpose' are reaped by economic elites while the affected people tend to belong to the most marginalised and economically weak strata of society (cf. Yerramilli, 2017). International tourism industry watchdogs, such as Tourism Concern, have argued that the notion that land can be acquired 'for public purpose' through tourism should be categorically refuted (IHRB & Tourism Concern, 2012). The justification of development-induced evictions (including for tourism and sporting events) under the pretext of serving the 'public good' has also been dismissed by the United Nations' Special Rapporteur on adequate housing (OHCHR, 1997; see also UN Habitat and OHCHR, 2014).

Another common narrative is the *'idle land' discourse* which suggests that the area to be developed for tourism purposes has no alternative value, as it is considered 'wasteland', 'degraded land' or 'underutilised land'. In such a discourse, the original land users are depicted as people that are not able to make productive and sustainable use of their land. Yet, in many cases, the landowners that have been dispossessed or evicted were previously engaged in highly profitable or at least self-sustaining livelihood activities. For instance, coastal dwellers displaced by a large-scale Chinese tourism projects in Cambodia's Koh Kong Province were successful cashew nut and rice growers prior to being relocated about 20km inland into the Botum Sakor National Park (see Chapter 3). Similarly, women in Rote Ndao (Eastern Indonesia) whose foreshore areas have been enclosed by a domestic resort development were previously involved in profitable seaweed farming (Chapter 4). Several other cases in this book also demonstrate how the myth of 'idle land' is perpetuated in government and corporate discourses. The case of Jerusalem's tourist map of 2016 in which a densely populated Palestinian settlement to the east of the city wall was falsely depicted by Israel's Ministry of Tourism as an uninhabited green landscape is one such example.

The *'tourism for conservation' discourse* is often used by national governments and international NGOs to justify the preservation of natural and cultural heritage sites through removing and replacing local people's previous livelihoods with ecotourism (e.g. the case of Colombia's Tayrona National Park in Chapter 8), wildlife tourism (cf. case studies from India in Chapter 7) and cultural tourism (e.g. the cases of Guatemala's Maya Biosphere Reserve and Cambodia's Angkor Archaeological Park). In such discourses, the original land-owners and customary users of the 'heritage space' are branded as 'eco-threats', as exemplified by the case of the Masaai pastoralists in Tanzania (Chapter 7) and the Garifuna communities in Honduras (Chapter 5). Alternatively, their previous roles in maintaining (eco-)cultural heritage are simply ignored or dismissed, as evidenced by the case of former caretakers of temples that were evicted from the Bagan cultural heritage sites in Myanmar (Chapter 6). The tourist sector, by contrast, is depicted in government and corporate discourses as environmentally and culturally benign and as a tool to maintain and restore the integrity of (eco-)cultural heritage sites and protected areas. Such discourses have also been supported by mainstream 'tourism and development' literature (e.g. Telfer and Sharpley, 2008) and international interest groups, such as the WTTC and UNWTO.

The *'tourism and security' discourse* is primarily employed in conflict and post-conflict contexts. The presence of military or paramilitary forces around sites of touristic interest is discursively justified by the assumed right of both cos-mopolitan and domestic tourists to a safe and carefree vacation. In Sri Lanka, for instance, this discourse has legitimised the ownership and control of a range of tourism facilities by the country's military and the continued presence of armed forces in places where domestic (Sinhalese) tourists are presented with a one-sided perspective of the long-standing civil war against the Liberation Tigers of Tamil Eelam (LTTE) and their defeat in 2009 (Seioghe, 2016; cf. Chapter 6). Similar official government discourses are at work in Bangladesh's Chittagong Hill Tracts where the military has exerted control over strategically important touristic infrastructure in Indigenous Jumma territories by dis-possessing and displacing local communities and providing domestic visitors with a safe and nationalist touristic experience (Chapter 6). Furthermore, in Jerusalem, the Israeli government advances its claims to territorial sovereignty over the eastern part of the city through a combination of spatial securitisation and cultural erasure, reinforced by discourses that leave no place for Palestinian interpretations of archaeological sites visited by Israeli and foreign tourists (Ezrahi and Mizrachi, 2020; Chapter 6).

Since the early 2000s, large-scale land acquisitions and displacement for a variety of purposes have often been justified by making recourse to a *'crisis' discourse*, particularly referring to the food, water, energy and climate crises (White et al., 2012; Neef, 2014). In the context of tourism-related land grabs, such crisis discourses have been deployed following major financial and eco-nomic crises, as evidenced by the case of the Cancún megaresort development in the midst of an economic crisis in Mexico in the early 1970s (Chapter 3) and aggressive tourism development in Costa Rica following a depression affecting the country's agricultural sector in the 1980s (Chapter 4). Similar crisis narratives were at play in the rapid move towards mass tourism in Bali, Indonesia, following the 1997 Asian financial crisis, when the sector was hailed as a motor of the country's recovery (Chapter 4). Crisis discourses that are used to legitimise unfettered tourism development and associated land grabs and displacement have been particularly prevalent in post-disaster contexts, e.g. in Thailand, Sri Lanka, India and the Maldives following the devastation of islands and coastal areas by the 2004 Indian Ocean Tsunami, in the aftermath of the Haiti Earthquake of 2010 and in the wake of cyclones and typhoons in the Asia-Pacific region, as exemplified by the case of Sicogon Island in the Philippines following 2016 super-typhoon Haiyan (Chapter 5).

Concluding remarks

This chapter has introduced the global tourism industry as a polymorphic assemblage of actors. The aspirations of governments in the Global South to expand tourism tend to converge with the interests of transnational tourism corporations that have little regard for the customary rights of local communities. Land grabs and displacement are driven by a confluence of excess capital and excessive mobilities and are supported by a range of discourses that are deployed by governments, corporate actors, bilateral donors and international financial institutions to legitimise such human rights violations. The involvement of armed forces and private security firms – particularly but not exclusively in authoritarian regimes – has contributed to both the implementation and securitisation of tourism-associated dispossession and displacement. Financialisation by sovereign wealth funds, pension funds, internationally operating commercial banks and private equity firms plays an indirect but crucial role in the proliferation of land grabs and evictions for tourism purposes. Tourism watchdogs, pressure groups and transnational social movements have emerged as an important counterbalance to the predatory activities of the tourism sector but are facing an uphill battle.

The next chapter will introduce the state as one of the key players in tourism-related land grabs and displacement in the Global South. Instruments employed by the state include the delineation of tourism zones, the provision of investor-friendly land legislation and economic incentives, and the direct allocation of land concessions at discounted rates to tourism investors. Case studies from Mexico, the Philippines, Cambodia and Timor-Leste will illustrate the use of these instruments by state actors, the alliances that state entities form with other powerful actors and the impact these have on the customary land rights of communities in the targeted areas.

References

Capapé, J. (2016) 'Sovereign wealth funds check-in: Investment strategies in the hotel sector' in Santiso, J. and Capapé, J. (eds) *Sovereign Wealth Funds 2016*. Tufts University: Boston, MA, pp. 55–66. Available at: https://sites.tufts.edu/sovereignet/files/2017/08/Report_Sovereign-Wealth-Funds-2016.pdf (accessed 14 August 2020).

Ezrahi, T. and Mizrachi, Y. (2020) 'Cable car plan threatens unique character and heritage of the Old City of Jerusalem' in World Heritage Watch (ed.) *World Heritage*

Watch Report 2020. Berlin, pp. 138–142. Available at: https://world-heritage-watch. org/wp-content/uploads/2020/06/WHW-Report-2020.pdf (accessed 12 July 2020).

Gibson, C. (2019) 'Critical tourism studies: New directions for volatile times', *Tourism Geographies*. doi:10.1080/14616688.2019.1647453.

Hall, R., Edelman, M., Borras, S.M. Jr., Scoones, I., White, B. and Wolford, W. (2015) 'Resistance, acquiescence or incorporation? An introduction to land grabbing and political reactions "from below"', *Journal of Peasant Studies* 42 (3–4): 467–488.

Harvey, D. (1989) *The Condition of Postmodernity*. Basil Blackwell: Oxford.

Hong, E. (1985) *See the Third World While It Lasts: The Social and Environmental Impact of Tourism with Special Reference to Malaysia*. Consumers' Association of Penang: Penang.

IHRB & Tourism Concern (2012) 'Frameworks for change: The tourism industry and human rights'. Available at: www.ihrb.org/pdf/2012-05-29-Frameworks-for-Cha nge-Tourism-and-Human-Rights-Meeting-Report.pdf (accessed 19 January 2019).

Mason, P. (2016) *Tourism Impacts, Planning and Management* (3rd edition). Routledge: London, New York.

Mowforth, M. and Munt, I. (2016) *Tourism and Sustainability: Development, Globalisation and New Tourism in the Third World* (4th edition). Routledge: London, New York.

Neef, A. (2014) 'Law and development implications of transnational land acquisitions: Introduction', *Law and Development Review*, 7 (2): 187–205.

OHCHR (1997) General comment No. 7: The right to adequate housing (art. 11 (1) of the Covenant): Forced evictions. Available at: https://tbinternet.ohchr.org/_la youts/treatybodyexternal/Download.aspx?symbolno=INT/CESCR/GEC/6430&La ng=en (accessed 24 January 2019).

PATA and WINTA (2015) Indigenous tourism & human rights in Asia & the Pacific Region: Review, analysis, & checklists. Available at: www.ecotourism.org.au/assets/ Resources-Hub-Indigenous-Tourism/International-Indigenous-Tourism-Human-Ri ghts-Review-Analysis-Checklists.pdf (accessed 21 December 2018).

Seoighe, R. (2016) 'Inscribing the victors land: Nationalistic authorship in Sri Lanka's postwar Northeast', *Conflict, Security and Development*, 16 (5): 443–471.

Swarbrooke, J. (1999) *Sustainable Tourism Management*. CABI Publications: Wallingford.

Telfer, D.J. and Sharpley, R. (2008) *Tourism and Development in the Developing World*. Routledge: London, New York.

UN Habitat and OHCHR (2014) 'Forced evictions – Fact Sheet No. 25'. Available at: www.ohchr.org/Documents/Publications/FS25.Rev.1.pdf (accessed 21 December 2018).

White, B., Borras Jr., S.M., Hall, R., Scoones, I. and Wolford, W. (2012) 'The new enclosures: Critical perspectives on corporate land deals', *The Journal of Peasant Studies*, 39 (3–4): 619–647.

Yerramilli, S. (2017) 'The "public purpose" that is not inclusive' in Pellissery, S., Davy, B. and Jacobs, H.M. (eds) *Land Policy in India*. Springer Nature: Singapore, pp. 127–145.

Zoomers, A. (2010) 'Globalisation and the foreignisation of space: Seven processes driving the current global land grab', *The Journal of Peasant Studies*, 37 (2): 429–447.

3 State-led tourism development, tourism zoning and customary land rights

As discussed in Chapter 2, one of the dominant discourses employed to promote tourism ventures is that of tourism as an important driver of economic development through the provision of jobs and as a foreign exchange earner for the host country. Therefore, many national governments in the Global South have prioritised the tourism sector as an engine of economic growth and developed top-down tourism strategies, such as national tourism plans, state-developed infrastructure and incentives for the development of tourism facilities (Richter, 2008). It is important to note that 'the state apparatus' does not only include elected officials but also more or less "permanent institutions and the elites that staff them" (Clancy, 2001, p. 17). The state's role in tourism development may include any combination of planning, provision, securitisation, risk taking, financing as well as expropriation and relocation of communities that stand in the way of tourism development.

The establishment of Mexico's Cancún megaresort in 1973/74 and the subsequent expansion of tourism along the Riviera Maya has probably been the most commonly cited case of state-led tourism development in the literature (Clancy, 2001; Murray, 2007). This case will be examined in the first section of this chapter. It will be followed by a discussion of recent state-led tourism development projects in the Philippines. This Southeast Asian country has an equally long-standing history of declaring huge tracts of lands and entire islands as tourism zones, going back to the early 1970s when then President Marcos used tourism for personal gains and political leverage (Richter, 1999). Another section in this chapter will examine the case of a tourism mega-concession leased by a Chinese state-owned company in Cambodia's southwestern Koh Kong Province. The final section will discuss how the government of Timor Leste has attempted to centrally develop tourism in its Oecusse enclave, as the country – one of the poorest in the Asia-Pacific region – faces a major economic downturn with the depletion of its offshore oil and gas reserves.

Foreignising paradise through state-led tourism development – the case of Cancún and La Riviera Maya, Mexico

Until the early 1970s, Mexico's tourism sector was small and informal, with only the coastal resort of Acapulco attracting a sizable number of foreign tourists, hailing mostly from the United States (Gladstone, 2005). Former Mexican President Miguel Alemán-Valdes (1946–1952) who later became the General Director of the National Tourism Commission for 25 years is commonly referred to as the 'father of the Mexican tourism industry'. This 'title' was owed to his development of strategic plans for tourism in the 1950s and his channelling of investments to Acapulco where he and his cronies had vested interests, as they had previously acquired large tracts of land in the area (Clancy, 2001; Ambrosie, 2015).

Back then, the north-eastern corner of the Yucatán peninsula in the Mexican state of Quintana Roo was one of the poorest and most remote regions of the country (Padilla, 2015). What would become one of the most famous international tourism hotspots – Cancún – was a small sandspit with an adjacent community of no more than 500 Indigenous Mayans who eked out their modest livelihoods as fishers and farmers. Suffering from the collapse in demand for sisal fibre and a nation-wide economic crisis in the early 1970s, the region was deemed in urgent need of an injection of national and foreign investment through absorbing US American tourists that had been 'exiled' from their favourite holiday spots in Cuba (Ambrosie, 2015). The creation of Integrally Planned Centres (IPCs) or tourism megaprojects was regarded by government planners and the Bank of Mexico as an ideal instrument to valorise underutilised natural resources in coastal areas (Vargas Martinez, Castillo Nechar and Viesca González, 2013). The choice of Cancún – its name derived from the Mayan word for 'nest of serpents' – as the first megaproject was reportedly a result of market analysis combined with computer-supported studies of weather patterns by bankers in the country's capital Mexico City (Clancy, 2001). The area was also considered sparsely populated and therefore provided few obstacles to the relocation of locals (Ambrosie, 2015).

The early phase of Cancún's development was managed by the Bank of Mexico's National Trust Fund for Tourism Infrastructure (*Fondo de Promoción de Infraestructura Turística* – INFRATUR). This fund – like its successor *Fondo Nacional para el Desarrollo Turístico* (FONATUR) – was endowed with the power to expropriate land and relocate communities for the purpose of tourism development (Ambrosie, 2015; cf. Box 3.1). To make way for the Cancún hotel zone and the adjacent 'service city' Ciudad Cancún, these agencies expropriated 170 Indigenous landowners living on the sandspit and its surrounding area, dredged lagoons and removed mangrove forests (Clancy, 2001). The area 'acquired' from local people constituted only about 400 hectares, while the remaining 7,340 hectares comprised a land grant under a presidential decree (Ambrosie, 2015). Funding for the development of the tourism infrastructure was provided by trust fund monies and the country's national development bank (NAFINSA) which received a large loan from the Inter-American Development Bank (IDB), its first direct financing of a tourism project (Clancy, 2001; Murray, 2007). By 1981, the IDB would have provided more than US$300 million for tourism development in Mexico, including two large additional loans for Cancún (Clancy, 2001).

Box 3.1 FONATUR: The world's most powerful national tourism organisation?

FONATUR (*Fondo Nacional para el Desarrollo Turístico*) is arguably one of the most powerful national tourism organisations in the world, overseeing an annual multi-billion US$ budget. It was founded in 1974 through a merger of the National Trust Fund for Tourism Infrastructure (*Fondo de Promoción de Infraestructura Turística* – INFRATUR) within the Bank of Mexico and a half-dormant financial trust fund *Fondo de Garantía y Fomento al Turismo* – FOGATUR.

FONATUR actively promotes foreign investment in tourism real estate by providing extremely favourable conditions for investors, including 100 per cent participation in shared capital and subsidised loans for hotel construction. FONATUR also provides and maintains the entire infrastructure around its resort developments, including airports, marinas, access roads, electricity networks, freshwater supplies and sewage systems.

Apart from Cancún, its most famous integrated resort development, FONATUR has developed at least four other integrated resort complexes, namely Los Cabos, Ixtapa, Loreto and Huatulcoy Bays. Other large-scale tourism infrastructure projects planned and implemented by FONATUR include the 333-berth Cozumel Marina and the 'Maya Train' megaproject (cf. Chapter 9). FONATUR operates with near-complete autonomy, can bypass local and provincial legislation, and has the power to expropriate land owners for tourism as a public benefit. In Ixtapa, for instance, large-scale expropriation of communal lands (*ejidos*) by FONATUR led to intense conflicts with the local Indigenous population.

FONATUR has also provided advisory services for tourism development in other Latin American countries, including Cuba and the Dominican Republic. The creator and first head of FONATUR, known as 'the father of Cancún', later became the Secretary General of the United Nations World Tourism Organization (UNWTO). Hence, FONATUR's influence on tourism development has extended well beyond Mexico's borders.

Sources: Clancy, 1999, 2001; Hiernaux-Nicolas, 2003;
Telfer and Sharpley, 2008; Ambrosie, 2015

One of the most significant hurdles to attracting foreign investment in support of continued tourism growth in Cancún was of a legal nature; the 1917 Constitution of Mexico did not allow foreigners to own property in coastal areas, i.e. land within 50 kilometres of any ocean (Murray, 2019). To get around this obstacle, FONATUR created a special type of trusteeship (*fideicomiso*) to regulate land transfers for the benefit of international investors (Clancy, 2001). Rather than taking the form of a lease, a *fideicomiso* is a renewable trust held initially for 50 years through a Mexican fiduciary bank that holds the title to the property on behalf of the foreign investor (Murray, 2019). Yet, initially, foreign investors remained hesitant to invest in property in a hitherto unknown tourist destination. In 1975, Club Méditerranée agreed to the operation of a 600-room hotel on a franchise basis, while INFRATUR/FONATUR had to take care of the construction and maintain ownership of the land and its premises (Clancy, 2001). International hotel chains only started to invest in hotel properties in the early 1980s, at which point Cancún's infrastructure was already well developed and tourist numbers started rising by double-digit percentages annually (Ambrosie, 2015).

In order to entice Mexican investors, INFRATUR/FONATUR offered discounts on the sale price of land, provided land-for-share swaps and invested in the megaresort's infrastructure (Ambrosie, 2015). As the megaresort was built from the ground up, FONATUR ultimately became the governing power of the area, with one of its executives even assuming the role of city mayor (Clancy, 2001). By the end of the 1980s, Cancún had become the single largest tourist destination in Mexico (Clancy, 1999) and in 2017, the resort city welcomed more than six million visitors (Wiedeman, 2019).

The rapid expansion of the hotel zone and the service city from the 1990s onward came at considerable environmental costs; while initially some land had been set aside for environmental protection (Clancy, 2001), hundreds of hectares of wetlands and mangrove forests were dredged and destroyed in violation of strict national environmental laws and the Ramsar Convention on Wetlands of International Importance to which Mexico had become a signatory in 1986 (Brundage, 2016). In 2015, local protests formed against another massive tourism development promoted by FONATUR, the Tajamar Malecón project, which threatened the existence of 143 acres (56 hectares) of mangrove forest and its unique wildlife (Mangrove Action Network, 2016). In a particularly cynical move, an injunction promoted by 113 local children – who invoked their right to a healthy environment – to suspend the project was granted by a local judge on the condition that the kids pay the equivalent of more than US$1 million in compensation to the foreign and domestic investors (Hernández, 2016). When the mangrove area was ultimately destroyed in a pre-dawn operation protected by strong riot police presence, FONATUR insisted that the action did not represent 'environmental damage' but only 'environmental impact' based on authorisation by the appropriate authorities (Brundage, 2016; Hernández, 2016).

Land invasions and resource grabs by domestic and foreign investors in Cancún have also proven detrimental to other rights of local residents. Following the allotment of private beach concessions in the 1990s, only two public beaches – Playa Delfines and Playa Langosta – remain that can be enjoyed by the inhabitants of Cancún (Vargas Martinez, Castillo Nechar and Viesca González, 2013). A comprehensive study by Domínguez Aguilar and García de Fuentes (2007) exposed the drastic inequality in freshwater supply in Cancún; while the hotel zone enjoys a piped water service without interruptions, households in the city centre receive piped water for only a few hours per day. Meanwhile, the poorest households in the peripheral and out-lying squatter settlements have neither provision for piped water nor sewerage and have to meet their water demand either through tanker services – at ten times the cost that hotel operators and most households in the city centre would pay – or from shallow wells that residents have to dig themselves (Aguilar and de Fuentes, 2007).

Notwithstanding the megaresort's significant contribution to Mexico's tourism revenues, Cancún has gone through a series of economic setbacks, primarily caused by natural and human-made disaster events. Among the most damaging ones – prior to the COVID-19 pandemic of 2020 – were Hurricanes Gilbert in 1988 and Wilma in 2005, the 9/11 terror attacks that dramatically reduced tourist numbers from the USA and the AH1N1 influenza pandemic of 2009 (Vargas Martinez, Castillo Nechar and Viesca González, 2013). Hurricane Wilma was particularly destructive, causing an estimated economic damage of about US$1.4 billion (Moncada, 2009). The impact of extreme storm events has been exacerbated by the removal of protective vegetation for hotel and infrastructure development, particularly the destruc-tion of coastal mangrove forests mentioned above. Hurricane Wilma and other major storms in combination with ocean currents have accelerated beach erosion, forcing the government to spend the equivalent of tens of millions of US dollars to replenish the beaches around Cancún (Lacey, 2009; *Mexican News Daily*, 2017). In 2010, for instance, more than five million cubic metres of sand were reportedly dredged from marine banks of sur-rounding islands and poured onto Cancún's beaches despite strong opposi-tion from environmental organisations (Vargas Martinez, Castillo Nechar and Viesca González, 2013).

Sand grabbing has also been common for the provision of construction material for the hotel infrastructure (Pérez Villegas and Carrascal, 2000). Seeing their most important asset erode into the ocean, some hotel owners have taken matters into their own hand and built artificial reefs and seawalls to halt beach erosion (Barrell, 2010). A hotel owner was fined and five of his employees detained in 2009 for building a breakwater into the ocean and illegally pumping sand from the ocean floor to the hotel beach (Lacey, 2009). In 2017, the tourism secretary of Quintana Roo ordered the suspension of 15 private beach recovery projects and announced a US$30 million beach reha-bilitation plan, with major investment from a Spanish construction company (*Mexico News Daily*, 2017). Yet it is only a matter of time until the next storm event will undo these desperate efforts. In fact, during the writing of this chapter, Hurricane Delta hit the Yucatán Peninsula just south of Cancún in early October 2020 but – fortunately – it was not as destructive as previous storm events.

Meanwhile, Cancún itself has been described as a 'hurricane' bringing mass tourism to other parts of the *Riviera Maya*, the name given to the 140km eastern Yucatan coastline running from Cancún down to Tulum. What has until recently been considered the "last, undeveloped Caribbean beach property in North America" (Houghton, 2012, n.p.) is being rapidly bought up by US American, Canadian and European investors. In 2006, Playa del Carmen, 70km south of Cancún, was considered one of the fastest growing cities in the world (Guy, 2018). The land rush – mostly driven by small-scale foreign investment – has reached as far as Tulum, formerly a sleepy fishing village, whose population tripled between 2008 and 2015 (Guy, 2018). Large international hotel chains have so far shunned the area due to insecure land tenure and conflicting property claims (Monroe, 2017). In the 1970s, the Mexican government designated around 10,000 hectares of land around Tulum as communal land *(ejido)* to landless peasants, many of them Indigenous Mayans (Wiedeman, 2019). When changes to the Mexican constitution and a new Agrarian Law in 1992 allowed the privatisation of *ejidos* (Farley et al., 2012), many *ejidatarios* (communal landholders) started to sell their plots to domestic and foreign investors, in some cases to several buyers. The resulting multiple claims to the same beachfront property led to a series of seizures of foreign-owned hotels in Tulum between 2002 and 2016 (Semple, 2016; Monroe, 2017; Wiedeman, 2019). More than 30 hotel owners were evicted under threat of violence, often supported by court orders, with the confiscated properties rapidly resold and redeveloped (Monroe, 2017; Wiedeman, 2019). Despite strong condemnation of the evictions by several foreign embassies, none of the evicted owners have been able to return (Monroe, 2017).

Tourism zone development in the Philippines

State-led tourism development in the Philippines can be traced back to the autocratic government of President Ferdinand Marcos (1965–1986) who created the cabinet-level Department of Tourism shortly after the declaration of martial law in 1972 (Richter, 1999). His militarised tourism strategy used the discourse of 'making the Philippines safe for tourists' and followed his earlier dedication of a Pacific War Memorial on Corregidor island in 1968 – a former battle site that played a role in the invasion and liberation of the Philippines from Japanese military forces in World War II – which subsequently became a site of remembrance for domestic visitors as well as American and Japanese tourists and war veterans (Gonzalez, 2013). The establishment of large five-star hotel complexes in strategic places across the country – most of them owned by Marcos and his cronies – were reportedly financed using the country's national pension system as a guarantee (Richter, 1999). Apart from enriching himself and his associates, Marcos used tourism for political leverage at a time when the US had important military bases in the country (Richter, 1999; Gonzalez, 2013).

Until the onset of the COVID-19 pandemic in early 2020, tourism was one of the largest foreign exchange earners in the Philippines, and consecutive administrations have pursued extremely proactive policy measures to attract investors into the tourism industry. The Tourism Act of 2009 established a new agency, the Tourism Infrastructure and Enterprise Zone Authority (TIEZA), under the Department of Tourism. TIEZA is in charge of promoting the establishment of Tourism Enterprise Zones (TEZs), which may fall under any of the following zone classifications: (i) cultural heritage; (ii) health and wellness; (iii) ecotourism; (iv) general leisure; and (v) mixed use tourism. TEZs need to have a size of at least 5ha and can be either brownfield (with existing infrastructure) or greenfield (with no or minimal infrastructure) developments. The minimum amount of investment is US$5 million, not including land acquisition costs.

The operators of TEZs benefit from a range of fiscal and non-fiscal incentives, including multi-year income tax holidays, exemptions from taxes and custom duties on imports of capital investments and equipment, and tax deductions on costs incurred by environmental protection efforts, cultural heritage preservation activities, and sustainable livelihood programmes for local communities. As of January 2019, TIEZA administered 11 flagship Tourism Enterprise Zones and several other TEZs, such as Hacienda Looc (discussed in the next section). These are complemented by Tourism Economic Zones which are administered by the Philippine Economic Zone Authority (PEZA). As of 30 November 2017, there were 19 Tourism Economic Zones across the archipelago with a total approved investment of about US$ 900 million. Nearly all of them were under Filipino corporate management. One of these Tourism Economic Zones is located on Boracay Island (see below).

Tourism enterprise zone development for the elite in Hacienda Looc, Batangas Province

The Hamilo Coast project in Hacienda Looc is located in the Nasugbu municipality of Batangas province and can be conveniently reached from the Philippine capital Manila. It is being implemented by SM Prime, a subsidiary of SM Investments Corporation (hereafter called SM), which is one of the largest business conglomerates in the country. Hamilo Coast's master plan is being developed by IMA Design, Inc., which prides itself in having designed such international megaprojects as Disneyland Paris and Universal Studios. With the planned development of a golf course, two marinas, a shopping mall, a ferry terminal, ten resorts and more than 3,000 dwelling units, its vision is to become the premier sustainable beach resort town of the Philippines (Atkins, 2016).

Hacienda Looc extends over an area of 8,650ha and comprises four village clusters (*barangays*) with a combined population of about 10,000 (Atkins, 2016). The area has a chequered legal history; it was once in the hands of a rich Filipino family which had to hand over the ownership of the large estate to a state-owned development bank in 1973 when it was unable to repay a loan. In the 1980s, the estate was turned over to the Assets Privatization Trust, and subsequently a major share of the area was placed under the

Comprehensive Agrarian Reform Program (CARP) (see Box 3.2), with several hundred farmers being the intended beneficiaries of land ownership certificates (APC and KMP, 2012). Yet, the suitability of Hacienda Looc for agricultural use was questioned by a government-commissioned feasibility study which recommended converting the area into a tourism zone (Atkins, 2016). In the mid-1990s, a subsidiary of SM won the public bidding for those areas not covered by CARP and filed petitions to the Department of Agrarian Reform to cancel the land ownership certificates for the remainder of the estate (APC and KMP, 2012; Atkins, 2016). In 2007, the area was declared a Tourism Enterprise Zone by executive order of then President Macapagal-Arroyo.

Box 3.2 Pro-poor and indigenous land policy programmes in the Philippines

The *Comprehensive Agrarian Reform Program* (CARP) was introduced by then President Corazon Aquino in 1988 after intense lobbying by peasant movements and land reform advocates. Intended beneficiaries were landless, including tenants and regular or seasonal farmworkers, and land-poor farmers owning no more than three hectares of agricultural land. Tools for redistribution included voluntary sales with a compensation premium, compulsory acquisition and distribution of stocks held in land-based enterprises. The program was met with fierce resistance by landlords, who challenged its legitimacy and tried to exploit various loopholes to prevent land redistribution and, in some cases, even to re-appropriate distributed land, e.g. by lease-back or 'joint-venture' arrangements. The slow pace of CARP has also allowed many landlords to forcibly remove tenants from their land in order to avoid having their land subjected to agrarian reform.

The Indigenous Peoples Rights Act (IPRA) of 1997 recognises the rights of Indigenous peoples to their cultural integrity and self-governance and certifies customary property rights to ancestral domains and lands. The Act requires a council of elders to formally represent the community in all dealings with government entities. The *Certificate of Ancestral Domain Title* (CADT) recognises indigenous communal land rights on formerly public domain land. The process of obtaining a CADT is expensive and time-consuming, due to the large amount of evidence that needs to be provided by the claimants. Besides, ancestral domain claims often overlap with land in the 'public domain', such as protected areas and government reservations, as well as concessions given for mining, tourism, logging, plantations, and energy projects, which are governed by other existing and often conflicting laws.

Sources: Hall, Hirsch and Li, 2011; Neef, 2016

From the mid-1990s until the early 2010s, local communities together with a range of human rights advocacy groups staged public protests, organised media campaigns and launched court appeals. The case also featured in a 40-minute documentary *The Golf War* by filmmakers Jen Schradie and Matt DeVries. In response to fierce local resistance, Hacienda Looc experienced increased militarisation through national armed forces alongside private security guards employed by SM (Atkins, 2016). An NGO network led by the Asian Peasant Coalition (APC) and KMP reported a range of human rights violations against the protesters, including fatal shootings, death threats, burning of houses, destruction of crops, and farmers being prevented from accessing their fields (APC and KMP, 2012). Local people also complained about increased soil erosion, pesticide contamination, flooding and landslides as a result of the construction of a golf course in the area (APC and KMP, 2012).

In an attempt to make the land grab appear more legitimate and to co-opt local people into supporting its mega-project, SM has deployed a range of strategies. Through its corporate foundation, it launched a skills development programme and a livelihood project in the *barangays* of Hacienda Looc and built such amenities as basketball courts and day-care centres (Atkins, 2016). The Hamilo Coast project has also engaged in a strategic partnership with WWF Philippines that includes the creation and management of three Marine Protected Areas (MPAs) along the coast and the rehabilitation of mangroves and hillside forests (WWF Philippines, 2017; *The Manila Times*, 2018). Yet these measures of 'corporate greenwashing' have imposed serious restrictions on local people's livelihoods. Much of the waters off Hamilo Coast are now inaccessible to fisherfolks depending on nearshore fisheries, and charcoal making in the adjacent hillside forests – formerly an important source of income for landless people – has been banned since 2009 (Atkins, 2016).

Since the mid-2010s, local resistance against the Hamilo Coast project has become weaker, and several villagers have sold their land to SM and assumed low-wage employment on the company's premises (Atkins, 2016). Yet some NGOs recently renewed their calls on the government to revoke the ownership rights of large, colonial-era estates, including Hacienda Looc, and to redistribute the land to farming communities (Barahan, 2017).

Tourism zoning and rights of Indigenous peoples on Boracay Island

Boracay – a tiny island of slightly over ten km^2 in the Western Visayas region – has long been one of the prime destinations for international tourists visiting the Philippines for a beach holiday. The island obtained global media coverage in 2018 when President Rodrigo Duterte announced a six-month closure for tourists, calling it a "cesspool" that was in urgent need of a thorough clean-up (Domingo and Placido, 2018). The tiny island had received two million visitors in 2017, extending by far its ecological carrying capacity; yet, only a year before the presidential order to close the island and its about

400 hotels and restaurants, the Municipal Government of Malay in charge of governing tourism on the island under the Local Government Code of 1991 had mulled over the declaration of at least three more special economic zones to attract foreign and domestic investors to the world-renowned island (*The Daily Guardian*, 2017). Several studies conducted in the early 2010s already warned that Boracay had reached its ecological limits (e.g. Ong, Storey and Minnery, 2011; Smith et al., 2011). A much earlier wake-up call – when a coliform outbreak was reported by the Department of Environment and Natural Resources – was also not heeded by the Philippine government and tourism operators (Trousdale, 1999; White and Rosales, 2003).

In 1978, during the Marcos dictatorship, the entire island had been declared a tourist zone (and marine reserve) and was placed under the management of the Philippine Tourism Authority by Proclamation No. 1801. The proclamation was a de facto dispossession of the Indigenous Ati, a semi-nomadic hunter-gatherer group, who are the original settlers of Boracay (Baleva, 2019). After years of unplanned development, the Boracay Development Master Plan was formulated in 1990 to foster a more sustainable tourism development on the island, but the plan was never implemented as the responsibility for managing the island was transferred to the Municipality of Malay under a nation-wide decentralisation programme in 1991 (Ong, Storey and Minnery, 2011). In the mid-2000s, the mandate over the island shifted back to the national level, but this process was later reversed.

By the end of the 1990s, tourism development and the increased demand for land had pushed the Indigenous Ati to the fringes of the island society and forced them to live scattered across the island. In 1999, in preparation for a presidential visit, the Ati were relocated to a one-hectare piece of land, the so-called Bolabog Settlement where they faced harassment by private claimants (Baleva, 2019). A proposal by the Department of Tourism for an Ati housing project that would also serve as a tourist attraction was rejected by the Ati Tribal Council on the ground that this would be dehumanising. The local government and the Department of Environment and Natural Resources proposed a relocation to the mainland which was also dismissed, as the Ati emphasised their close ties to their ancestral land (Baleva, 2019). An offer of a one-hectare land donation made by a private real estate corporation in 2005 also failed, as the parties could not agree on the terms. Meanwhile, a proclamation in 2006 by then President Macapagal-Arroyo – which classified Boracay into 400 hectares of reserved forest land and close to 630 hectares of alienable and disposable agricultural land – caused widespread confusion among land claimants on the island, particularly from the tourism sector (Tenefrancia, 2010).

In 2008/09, the National Commission on Indigenous Peoples (NCIP) attempted to obtain a presidential land donation of about 2.1 hectares of protected forest on behalf of the Ati community but were unsuccessful due to opposition from the Department of Environment and Natural Resources (DENR). The Bolabog Ati Tribal Council had already submitted petition letters to the NCIP for the issuance of a Certificate of Ancestral Domain Title (CADT) (cf. Box 3.1) as early as 2000 and again in 2003 and 2006, but the process of consultation and data gathering took several years and was riddled with a series of delays and strong opposition from other government entities and private land claimants (Baleva, 2019). Eventually, the CADT was issued in January 2011 but the Ati could not take immediate possession of their 2.1 hectares of ancestral domain because the private claimants had illegally occupied the land. The Ati community began a gradual process of self-installation from April 2012 amidst harassments and threats of violence which culminated in the assassination of one of their spokespeople by a security guard of a hotel chain in February 2013 (Ranada, 2014). Subsequently, the Ati community faced multiple litigation cases relating to their ancestral domain (Baleva, 2019).

The unprecedented 2018 closure of Boracay Island by President Duterte appears to have led to a complete overhaul of the island's land legislation, including the legal status and the development prospects of the Ati community. Halfway through the six-month closure, President Duterte announced unexpectedly that the entire island would be placed under land reform and that he would return the land to its 'rightful native owners' (Ranada, 2018a). A proposal was made by the Agricultural Secretary to turn the Ati's ancestral domain into an 'agro-tourism area' with "a vegetable production area using a solar-powered greenhouse and dairy goat milk parlor" and "an organic restaurant where indigenous food recipes could be offered to tourists" (Gomez, 2018, n.p.). This top-down project proposal under the Department of Agriculture's 'Livelihood and Progress of Filipino Indigenous Peoples' programme bears an odd resemblance with earlier suggestions by the Department of Tourism to turn the Ati themselves into a tourist attraction. Another proposal – this time by the Department of Agrarian Reform – suggested that the indigenous Ati people should become coconut farmers after having qualified as land reform beneficiaries (Rivas, 2018).

Yet, shortly before the re-opening of the island to tourists, the Environment Undersecretary announced that only about eight hectares of land were available for distribution to the Ati for agricultural purposes (Ranada, 2018b). Nevertheless, President Duterte travelled to Boracay in November 2018 to personally distribute six Certificates of Land Ownership Award (CLOA) to 45 members of the Boracay Ati Tribal Organization (Domingo and Placido, 2018; Gita, 2018). Duterte alluded to the Ati's rights to sell this land off to "big businesses" on the island – contradicting his earlier statement that he would award the land to the Ati to prevent "those with money [from taking] the land and build resorts" (Ranada, 2018a, n.p.).

Both cases discussed above – Hacienda Looc and Boracay Island – show how public and private actors in the Philippines are interwoven in a complex and pluralistic legislative framework pertaining to tourism development and land rights, which can have detrimental impacts on marginalised rural communities, both Indigenous and non-indigenous.

Concessional tourism development: a Chinese mega-project in Koh Kong Province, Cambodia

As a post-conflict and post-socialist country, Cambodia entered the global tourism arena at a relatively late stage but is trying hard to catch up with other countries in the Southeast Asian region, such as the Philippines, Indonesia and Thailand. Its 2016–2022 Tourism Strategic Directions Plan focused strongly on achieving numerical targets, with little emphasis on sustainable and community-based tourism (Sofield, 2020). Until recently, Cambodia's tourism industry was concentrated around the Angkor Archaeological Park in Siem Reap Province (discussed in Chapter 8) and the coastal resort town of Sihanoukville which has seen massive Russian and Chinese investments, including the establishment of about 70 casinos. In an attempt to cash in on the Southeast Asian tourism boom, Cambodia's government has since focused on converting hitherto pristine coastal areas into new tourism hotspots. Koh Kong Province is located in the southwestern part of Cambodia and plays a major role for biodiversity and wildlife conservation as it accounts for 568,450 hectares out of 3.3 million hectares of the country's total protected land area. Botum Sakor National Park extends over 171,250 hectares of land, with more than 50 per cent of its borders stretching along the Cambodian coast (Neef and Touch, 2015).

In May 2008, the Royal Cambodian Government, represented by the Minister of Environment, signed a long-term lease contract of 99 years with the Chinese corporation Tianjin Union Development Group Company Ltd (TUDG) for the construction of a large-scale commercial and tourism development zone, dubbed 'Angkor Wat of the Sea'. The project covers more than 36,000 hectares in two districts (Kiri Sakor and Botum Sakor) and involved investment capital of about US$3.8 billion. The leased land infringed on 12 villages in five communes and a large portion of the two districts' coastal areas, giving the concession effective control over about 20 per cent of Cambodia's coastline. In August 2011, the government issued a sub-decree to reclassify an additional 9,100 hectares as a 'sustainable use zone' and granted a second land concession to TUDG to develop a water reservoir and hydropower plant. This meant that the overall size of the concession exceeded the legally allowed maximum of 10,000 hectares by more than four times. In order to circumvent the legal limits on foreign concession leases, the company temporarily disguised as a Cambodian-owned company. The original plans of the mega-project included a high-end tropical resort, landscaped luxury villas, a large golf course, a casino, a large water entertainment centre, an island park, and a cultural entertainment zone (Sao, 2015). Supporting infrastructure to be built include an airport and a deep-sea port.

The villages affected by the project had occupied this location for generations, deriving their livelihoods primarily from fishing, paddy rice farming and cashew nut production. According to the district officials involved in the assessment of the occupied land, the project affected a total of 1,163 families. The affected communities were reportedly not consulted about the project and its potential impacts but had noticed company representatives and government officials travelling through their communes and measuring land before the signing of the contract in 2008. The communities were officially informed of the project for the first time during a visit in November 2009 to Kiri Sakor district by officials from the Ministry of the Environment, Royal Cambodia Armed Forces, the Koh Kong provincial government and representatives of TUDG (Neef and Touch, 2015). Government officials informed them that they were settling on state public land and were therefore legally obliged to relocate (Drbohlav and Hejkrlik, 2018). During this visit, villagers were promised compensation land in the area where they were to be relocated. In addition to the 1,163 families, at least one primary school and three Buddhist pagodas needed to be relocated (Neef and Touch, 2015).

Negotiations for compensation packages were conducted in 2010, and about 1,000 families were resettled in 2011 (Drbohlav and Hejkrlik, 2018). Many relocated families reported that they had received less compensation than what had previously been negotiated. Some families resisted relocation and had their houses knocked down by the company's security guards (Neef, 2016; Figures 3.1 and 3.2). The area was also patrolled by a unit of the Cambodian Prime Minister's own bodyguards to squash any remaining pockets of resistance (Neef and Touch, 2015). A woman who tried to plant rice on the land from which she had been evicted was beaten up by the company's own security guards in June 2014 (Sen, 2014). A year later, a protest camp set up by some of

Figure 3.1 Abandoned coastal settlement after involuntary relocation

Figure 3.2 One of the houses that were knocked down by the company's security guards

the remaining residents was violently dissolved, while one of the female protest leaders attended an international conference in Chiang Mai, Thailand.

The relocation site spreads over around 3,600 hectares of land located deep inside the Botum Sakor National Park at a distance of about 20km from the coast. According to local authorities, the government issued a sub-decree to excise this forested area from the park for this purpose (Neef and Touch, 2015). TUDG assumed responsibility to provide basic infrastructure for the resettlement area, including schools, health centres and water wells. However, the health centre lacks qualified staff, and many of the water wells constructed by TUDG run dry during the extended dry season, while the roads and basic wooden houses provided by TUDG were also of very poor quality (Drbohlav and Hejkrlik, 2018). Seven years after the relocation, many houses have been partially damaged and numerous houses have been vacated (Figure 3.3), as many villagers decided to leave the area due to lack of viable livelihood opportunities. Village roads are often impassable during the monsoon season (Figure 3.4), compromising villagers' access to markets and health care and students' access to schools.

Many of the relocated families stated that their allocated farmland was not suitable for farming and for some households the fields were as far as five to six kilometres away from their homestead (Drbohlav and Hejkrlik, 2018). Damage to crops by wild elephants and other wildlife has also been a common problem. Reportedly, TUDG employees have made successful attempts to acquire

Figure 3.3 One of many abandoned houses in the resettlement site

Figure 3.4 Road condition in the relocation site during the monsoon season

land in the resettlement site, thus capitalising on the financial difficulties of the resettled households (Neef, 2019). Due to the lack of agricultural productivity and other livelihood opportunities, many villagers are now engaged in illegal logging. They have to pay bribes to local officials who turn a blind eye on the illicit activity. The timber is sold to Vietnamese or Chinese middlemen who organise transportation of the logs to Vietnam (Neef, 2019).

By 2017, one resort and a golf course under the name 'Dara Sakor Seashore Resort' were completed (Figure 3.5), along with a four-lane connecting highway and two large water reservoirs. Yet, during a visit by the author to the site in late 2017, the entire resort area appeared to have been abandoned and a near-complete casino – originally designed as the tourism complex's centrepiece – was deserted and showed early signs of decay (Figure 3.6). A US-based think-tank has alleged that the megaproject now serves a more geostrategic purpose; in fact, it was officially incorporated into China's Belt and Road Initiative in 2016 by Chinese President Xi Jinping (Thorne and Spevack, 2017). In March 2017, the China Development Bank (CDB), a financial institution under jurisdiction of the State Council, announced that it had underwritten a US$15 million 'Belt and Road' bond to support TUDG's building of a holiday resort on Cambodia's coast (*Stabroek News*, 2018).

Allegations have been raised that the megaproject has been turned into a 'Chinese colony' and could even be part of a wider scheme of the Chinese government to establish naval bases throughout the region (Thorne and

Figure 3.5 Partial view of the 18-hole golf course at the Dara Sakor Seashore Resort

Figure 3.6 The decaying casino at the Dara Sakor Seashore Resort

Spevack, 2017; Nachemson, 2018). Such plans have been denied by the Cambodian government as 'fake news', although it has been confirmed that the construction of a port and airport are underway (Chheng, 2018). Regardless of whether Chinese intentions are purely economic or pursue a wider geostrategic purpose, there are well-grounded fears that Cambodian citizens are denied the benefits from coastal industries, such as fishing and tourism.

The special economic zone of social market economy in Oecusse, Timor-Leste

Timor-Leste is one of the world's youngest countries and also one of its poorest. The largely agrarian nation is heavily dependent on foreign aid and its dwindling offshore oil and natural gas reserves. The country's tourism sector is still in its infancy but earmarked by the government as a future growth industry (Bouchon, 2020). Tourism development efforts have particularly focused on Oecusse-Ambeno, a coastal enclave located in the western part of Timor Island and separated from the rest of the country by West Timor, which forms part of Indonesia's Nusa Tenggara Timur Province. After the vote for independence from Indonesia in a UN-backed referendum in 1999, the enclave with a population of around 70,000 Meto Indigenous people was ravaged by an Indonesian Army supported militia which destroyed most of the enclave's infrastructure and displaced the majority of its Indigenous population (Rose, 2017).

In early 2013, the government of Timor-Leste announced that Oecusse was to be developed into a tourism and investment hub under the official name *Zonas Especiais de Economia Social de Mercado - ZEESM* (Special Economic Zone of Social Market Economy), with an earmarked budget of more than US $4 billion over a period of 20 years (Meitzner Yoder, 2015). One of the major drawcards for tourists − aside from its pristine beaches and high-potential diving spots − is Oecusse's historical status as the first landing spot of the Portuguese, hence first results of the massive economic overhaul were to be expected by the time of the 500-year commemoration celebrations in November 2015. The high-profile character of the megaproject was underscored by the appointment of a former prime minister as its head and major champion. During public information sessions that were organised from 2013 to 2014 primarily in the Timor-Leste capital Dili as well as on brief tours throughout the predominantly rural district of Oecusse, futuristic landscape visions and design diagrams were presented that had been developed by Portuguese development consultants and South Korean waterfront development firms (Meitzner Yoder, 2015).

It was obvious from the presentations by foreign consultants and government officials that the characteristics that determining whether a rural area would be designated as a tourism region, potential industrial processing site, or an agro-export production zone were based purely on scientific and technical factors (Meitzner Yoder, 2015). The plans did not consider any socio-cultural aspects, including pre-existing legal claims to the land by the Indigenous Meto people that are the dominant ethnic group in the area (Rose, 2017). Two planned resort enclaves − featuring various hotels, golf courses and a waterpark − would be located on smallholder rice fields owned and cultivated by a large number of smallholder families and would divert water from an ecologically fragile, water-scarce region (Meitzner Yoder, 2015).

By mid-2016, construction of a bridge, a port, an international airport, a luxury resort, a historic monument park and a special clinic were in progress, while the district's main road was also being reconstructed (Rose, 2017; Meitzner Yoder, 2018). These infrastructure projects have consumed around US$ 500 million in public funds, mostly derived from the country's petroleum fund − a sovereign wealth fund fed by revenues from Timor-Leste's oil and gas fields that are expected to be depleted within the next few years (Davidson, 2017). While authorities claim that 70 per cent of the workers involved in infrastructure development are local, this figure is contested by a Dili-based NGO which suggests that virtually all workers at the Suai airport construction site are Indonesian (La'o Hamutuk, 2017).

The largest impact has been felt by local residents on privately owned land along the road construction site who saw their fruit trees, wells, gardens and houses demolished with little warning and were forced to live in makeshift shelters, as the delivery of funds and materials for rebuilding homes was delayed (Rose, 2017). The cash payments that were eventually handed out only covered between 15 and 25 per cent of the cost for the construction labour (Meitzner Yoder, 2016). In another area, a hamlet of about 12 huts was completely removed to make way for a luxury hotel, with one technician being quoted as saying that these people were there "illegally" (Rose, 2017, p. 205). The majority of roadside families lost hundreds of dollars in timber and fruit from felled trees without receiving any compensation for these losses (Meitzner Yoder, 2016). Traditionally, trees in home yards have been used by the Meto people to mark their ownership claims for a particular area (Meitzner Yoder, 2011).

NGO representatives have criticised the government-appointed ZEESM Authority enacting their own policies, regulations and executive orders for the enclave, including managing state-owned property and deciding on land use conversions and expropriations (La'o Hamutuk, 2017). In fact, legislation promulgated in 2014/15 effectively excised Oecusse from national governance structures and fiscal processes (Meitzner Yoder, 2015). Both local NGOs and foreign academics have accused the ZEESM for focusing on promising hypothetical luxury tourists and use 'aspirational distractions of megaprojects' rather than building on the strengths and priorities of local communities (Meitzner Yoder, 2015; La'o Hamutuk, 2017). A World Bank report also questions the tourism potential of the area, referring to the absence of "world-class beaches" or other major tourist attractions and to insufficient "access to land" as major constraints (World Bank, 2016, p. 24). Meanwhile, throughout lowland Oecussi, residents continue to be extremely insecure about the rights to their remaining land and natural resources and the future viability of their livelihoods (Rose, 2017).

Concluding remarks

This chapter has introduced the state as one of the central actors in grabbing land and displacing people to make way for large-scale tourism developments. Tourism development is increasingly prioritised by governments in the Global South over environmental concerns and the legitimate land and resource rights of local people. As the examples from Mexico, Philippines, Cambodia and Timor Leste have shown, the state apparatus – which may include powerful national tourism organisations, national banks, and other agencies – deploys a range of mechanisms to expropriate and evict communities, e.g. through tourism zoning, allocation of concessions, government-commissioned feasibility studies, police and military force, threat of violence, and development of infrastructure for 'public purpose'. Community resistance to such state-led tourism development has often been squashed by the use of state violence.

The following chapter will show how state-led tourism development interacts with corporate forces and gradually expands from resort tourism to residential tourism whereby the rights of local citizens to a clean environment, access to beaches and near-shore fishing grounds, and access to safe drinking water are increasingly compromised. The chapter will draw on examples from Indonesia, Vanuatu, Costa Rica and Mauritius.

References

Ambrosie, L.M. (2015) 'Myths of tourism institutionalization and Cancún', *Annals of Tourism Research*, 54: 65–83.

Asian Peasant Coalition (APC) and Kilusang Magbubukid Ng Pilipinas (KMP) (2012) *Report of the International Solidarity Mission to Save Hacienda Looc and Peasant Livelihoods: Executive Summary*. Available at: www.democraciaycooperacion.net/IMG/p df/ISM-HL_Executive_Summary_Report.pdf (accessed 6 January 2019).

Atkins, S. (2016) *Land Grabbing, Tourism and Conservation in the Philippines*. Lambert Academic Publishing: Saarbrücken, Germany.

Baleva, M.K.A. (2019) *Regaining Paradise Lost: Indigenous Land Rights and Tourism*. Brill: Boston.

Bouchon, F. (2020) 'Can tourism be a success story? Stakeholders' management and narratives of rural tourism. Reflexive analysis of a tourism project in Timor Leste' in Nair, V., Hamzah, A. and Musa, G. (eds) *Responsible Rural Tourism in Asia*. Channel View Publications: Bristol and Blue Ridge Summit, pp. 177–194.

Calvan, D. (2015) *Land, Property and Tenurial Rights in a Changing Coastal Environment*. Land Watch Asia, Asian NGO Coalition for Agrarian Reform and Rural Development. Available at: www.angoc.org/wp-content/uploads/2015/12/Land-Prop erty-etc_final.pdfc (accessed 1 January 2019).

Clancy, M.J. (1999) 'Tourism and development: Evidence from Mexico', *Annals of Tourism Research*, 26 (1): 1–20.

Clancy, M. (2001) *Exporting Paradise: Tourism and Development in Mexico*. Pergamon: Amsterdam.

Domínguez Aguilar, M. and García de Fuentes, A. (2007) 'Barriers to achieving the water and sanitation-related Millennium Development Goals in Cancún, Mexico at the beginning of the twenty-first century', *Environment & Urbanization*, 19 (1): 243–260.

Drbohlav, P. and Hejkrlik, J. (2018) 'Social and economic impacts of land concessions on rural communities of Cambodia: Case study of Botum Sakor National Park', *International Journal of Asia Pacific Studies*, 14 (1): 165–189.

Export-Import Bank of China (2018) *Annual Report 2017*. Available at: http://english. eximbank.gov.cn/upload/accessory/201812/201812415543010122.pdf (accessed 14 January 2019).

Farley, K.A., Ojeda-Revah, L., Atkinson, E.E., Ricardo Eaton-González, B. and Gonzalez, V.V. (2012) 'Changes in land use, land tenure, and landscape fragmentation in the Tijuana River Watershed following reform of the ejido sector', *Land Use Policy*, 29: 187–197.

Gladstone, D.L. (2005) *From Pilgrimage to Package Tour: Travel and Tourism in the Third World*. Routledge: London, New York.

Gonzalez, V.V. (2013) *Securing Paradise: Tourism and Militarism in Hawai'i and the Philippines*. Duke University Press: Durham, London.

Hall, D., Hirsch, P. and Li, T.M. (2011) *Powers of Exclusion: Land Dilemmas in Southeast Asia*. University of Hawaii Press: Honolulu.

Hiernaux-Nicolas, D. (2003) 'Mexico: Tensions in the Fordist model of tourism development' in Hoffman, L.M., Fainstein, S.S. and Judd, D.R. (eds) *Cities and Visitors: Regulating People, Markets and City Space*. Blackwell Publishing: Malden, MA, Oxford and Carlton, pp. 187–199.

La'o Hamutuk (2017) *Special Economic Zone in Oecusse*. Available at: www.laohamutuk. org/econ/Oecussi/ZEESMIndex.htm (accessed 6 January 2019).

Meitzner Yoder, L.S. (2011) '*Tensions of tradition: Making and remaking claims to land in the Oecusse enclave*' in McWilliam, A. and Traube, E.G. (eds) *Land and Life in Timor Leste: Ethnographic Essays*. Australian National University Press: Canberra, pp. 187–216.

Meitzner Yoder, L.S. (2015) 'The development eraser: Fantastical schemes, aspirational distractions and high modern mega events in the Oecusse enclave, Timor-Leste', *Journal of Political Ecology*, 22: 299–321.

Meitzner Yoder, L.S. (2016) 'The formation and remarkable persistence of the Oecusse-Ambeno enclave, Timor', *Journal of Southeast Asian Studies*, 47 (2), 281–303.

Meitzner Yoder, L.S. (2018) 'Economic techno-politics and technocratic development in the Oecusse-Ambeno enclave, Timor-Leste', *The Asia Pacific Journal of Anthropology*, 19 (5): 395–411.

Moncada, P. (2009) 'Desastres y turismo [Disasters and tourism]', *Boletín Turístico de Cancún* [Tourism Bulletin of Cancún], 2 (2): 2–14.

Murray, G. (2007) 'Constructing paradise: the impacts of big tourism in the Mexican coastal zone', *Coastal Management*, 35 (2–3): 339–355.

Neef, A. (2016) 'Cambodia: Land grabs and dispossession by government design. *Rural*, 213: 20–22.

Neef, A. and Touch, S. (2015) '*Local responses to land grabbing and displacement in rural Cambodia*' in Price, S. and Singer, J. (eds) *Global Implications of Development, Climate Change and Disasters: Responses to Displacement from Asia–Pacific*. Routledge: London, New York, pp. 124–141.

Neef, J. (2019) *Livelihood Changes of Resettled Communities in the Koh Kong Province, Cambodia*. Unpublished Master thesis, University of Hohenheim, Stuttgart.

Ong, L.T.J., Storey, D. and Minnery, J. (2011) 'Beyond the beach: Balancing environmental and socio-cultural sustainability in Boracay, the Philippines', *Tourism Geographies*, 13 (4): 549–569.

Padilla, N.S. (2015) 'The environmental effects of tourism in Cancun, Mexico', *International Journal of Environmental Sciences*, 6 (1), 282–293.

Pérez Villegas, G. and Carrascal, E. (2010) 'El desarrollo turístico en Cancún, Quintana Roo y sus consecuencias sobre la cubierta vegetal [Tourist development in Cancún, Quintana Roo and its consequences on vegetation coverage]', *Investigaciones Geográficas, Boletín del Instituto de Geografía, UNAM*, 43: 145–166.

Richter, L.K. (1999) 'After political turmoil: the lessons of rebuilding tourism in three Asian countries', *Journal of Travel Research*, 38: 41–45.

Richter, L. (2008) '*Tourism policy-making in Southeast Asia: A twenty-first century perspective*' in Hitchcock, M., King, V.T., and Parnwell, M. (eds) *Tourism in Southeast Asia: Challenges and New Directions*. Northern Institute of Asian Studies Press: Copenhagen, pp. 132–145.

Rose, M. (2017) '"Development", resistance and the geographies of affect in Oecussi: Timor-Leste's Special Economic Zone (ZEESM)', *Singapore Journal of Tropical Geography*, 38: 201–215.

Sao, V. (2015) *A Study on Resettlement Schemes of Large Scale Land Lease to Chinese Investment in Cambodia: Case Study of Union Development Group, Co., Ltd*. Working Paper, Mekong Institute, Khon Kaen: Thailand.

Smith, R.A., Henderson, J.C., Chong, V., Tay, C. and Jingwen, Y. (2011) 'The development and management of beach resorts: Boracay Island, the Philippines', *Asia Pacific Journal of Tourism Research*, 16 (2): 229–245.

Sofield, T.H.B. (2020) '"MlupBaitong" – A pioneer in responsible rural tourism in Cambodia' in Nair, V., Hamzah, A. and Musa, G. (eds) *Responsible Rural Tourism in Asia*. Channel View Publications: Bristol and Blue Ridge Summit, pp. 162–176.

Telfer, D.J. and Sharpley, R. (2008) *Tourism and Development in the Developing World.* Routledge: London, New York.

Tenefrancia, R. (2010) *Boracay Island: A Case for Reversing Island Tourism Over-Development to Promote Sustainable Tourism.* Available at: https://pinoygreenacademy.typepad. com/files/white-paper_july6_final.pdf (accessed 12 January 2019)

Thorne, D. and Spevack, B. (2017) *Harbored Ambitions: How China's Port Investments Are Strategically Reshaping the Indo-Pacific.* C4ADS: Washington, DC.

Trousdale, W.J. (1999) 'Governance in context: Boracay Island', *Annals of Tourism Research,* 26 (4): 840–867.

Vargas Martinez, E.E., Castillo Nechar, M. and Viesca González, F.C. (2013) 'Ending a touristic destination in four decades: Cancun's creation, peak and agony', *International Journal of Humanities and Social Sciences,* 3 (8): 16–26.

Walker, C.J. and Curet, L.A. (2009) *Heritage or Heresy: Archaeology and Culture on the Maya Riviera.* University of Alabama Press: Tuscaloosa.

White, A.T. and Rosales, R. (2003) 'Community-oriented marine tourism in the Philippines: Role in economic development and conservation' in Gössling, S. (ed.) *Tourism and Development in Tropical Islands.* Political Ecology Perspectives. Edward Elgar: Cheltenham, Northampton, MA, pp. 237–262.

World Bank (2016) *Democratic Republic of Timor-Leste Oecusse Economic and Trade potential.* World Bank: Washington, DC.

Media sources/websites

Barahan, M. (2017) 'Include big landholdings, haciendas in free land distribution–KMP', *Inquirer.net,* 19 April. Available at: https://newsinfo.inquirer.net/890350/include-big-landholdings-haciendas-in-free-land-distribution-kmp (accessed 6 January 2019).

Barrell, S. (2010) 'The battle for the beaches of Cancun', *The Independent* (UK), 9 May. Available at: www.independent.co.uk/travel/americas/the-battle-for-the-beaches-of -cancun-1968917.html (accessed 4 June 2020).

Brundage, J. (2016) '"No environmental damage" to Tajamar Mangrove Cancún: Project to proceed – Mexico government', 27 January. Available at: https://voicesmotherea rth.blogspot.com/2016/01/no-environmental-damage-to-tajamar.html (accessed 4 June 2020).

Chheng, N. (2018) 'Government denies Koh Kong Chinese naval base "rumour"', *Phnom Penh Post,* 19 November. Available at: www.phnompenhpost.com/national-p olitics/government-denies-koh-kong-chinese-naval-base-rumour (accessed 1 January 2019).

Davidson, H. (2017) 'Timor-Leste's big spending: a brave way to tackle economic crisis or just reckless?', *The Guardian,* 25 May. Available at: www.theguardian.com/world/ 2017/may/25/timor-leste-spending-big-economic-crisis (accessed 6 January 2019).

Domingo, K. and Placido, D. (2018) 'Duterte distributes land to Ati tribe in Boracay', *ABS-CBN News,* 11 August. Available at: https://news.abs-cbn.com/news/11/08/ 18/duterte-distributes-land-to-ati-tribe-in-boracay (accessed 5 January 2019).

Gita, R.A. (2018) 'Duterte flies to Boracay to distribute land titles', *SunStar Philippines,* 8 November. Available at: www.sunstar.com.ph/article/1773152 (accessed 13 January 2019).

Guy, J. (2018) 'Uncontrolled development turns Mexican tourist paradise into an environmental time bomb', *Equal Times,* 18 January. Available at: www.equaltimes.org/ uncontrolled-development-turns?lang=en#.XxfhVCgzaUk (accessed 8 June 2020).

Gomez, E.J. (2018) 'Ati tribal land turned into Boracay agro-tourism area', *The Manila Times*, 28 June. Available at: www.manilatimes.net/ati-tribal-land-turned-into-bora cay-agro-tourism-area/413226/ (accessed 20 December 2018).

Hernández, A. (2016) '"Malecón Tajamar": Economic progress or environmental pitfall?' *Yucatan Times*, 22 January. Available at: www.theyucatantimes.com/2016/ 01/malecon-tajamar-economic-progress-or-environmental-pitfall/ (accessed 6 June 2020).

Houghton, R. (2012) 'Americans stake claim in Mexico's (Riviera Maya) land rush'. 12 September. Available at: https://blog.investmentpropertiesmexico.com/investment/ 2012/09/13/americans-stake-claim-in-mexicos-riviera-maya-land-rush (accessed 4 June 2020).

Lacey, M. (2009) 'A battle as the tide takes away Cancún sand', *New York Times*, 17 August. Available at: www.nytimes.com/2009/08/18/world/americas/18cancun. html (accessed 4 June 2020).

Lopez-Mills, D. (2010) 'Cancun's famous beaches threatened by erosion, climate change', *Associated Press*, 12 April. Available at: http://usatoday30.usatoday.com/travel/destina tions/2010-12-04-cancun-beaches-erosion-climate-change_N.htm (accessed 4 June 2020).

Mangrove Action Network (2016) 'Save the Mangroves at Tajamar', 10 June. Available at: https://mangroveactionproject.blogspot.com/2016/06/save-mangroves-at-tajama r.htm (accessed 4 June 2020).

Mexico News Daily (2017) '(Another) plan for the recovery of beaches', 25 February. Available at: https://mexiconewsdaily.com/news/another-plan-for-the-recovery-of-qr-beaches/ (accessed 4 June 2020).

Monroe, R. (2017) 'How rich hippies and developers went to war over Instagram's favourite beach', *The Guardian*, 26 April. Available at: www.theguardian.com/news/ 2017/apr/26/tulum-mexico-hotel-evictions-instagram-favourite-beach (accessed 4 June 2020).

Murray, H. (2019) 'What is the fideicomiso?' Available at: www.expatsinmexico.com/ what-is-the-fideicomiso/ (accessed 2 June 2020).

Nachemson, A. (2018) 'A Chinese colony takes shape in Cambodia', *Asia Times*, 5 June. Available at: www.atimes.com/article/a-chinese-colony-takes-shape-in-cambodia/ (accessed 16 January 2019).

Ranada, P. (2014) 'Violence looms over Ati tribe ancestral domain in Boracay', *Rappler*, 26 February. Available at: www.rappler.com/nation/51635-ati-tribe-security-threa t-ancestral-domain#cxrecs_s (accessed 19 December 2018).

Ranada, P. (2018a) 'Boracay natives can sell land to big firms, if they want – Duterte', *Rappler*, 13 June (updated 16 October). Available at: www.rappler.com/nation/2047 96-duterte-boracay-natives-sell-land-big-businesses#cxrecs_s (accessed 19 December 2018).

Ranada, P. (2018b) 'Only 8 hectares of Boracay land to be distributed to Ati tribe', *Rappler*, 23 October. Available at: www.rappler.com/nation/214985-only-8-hecta res-boracay-land-to-be-distributed-to-ati-tribe (accessed 23 December 2018).

Rivas, R. (2018) 'Boracay to become coconut producer under agrarian reform', 4 June. Available at: www.rappler.com/nation/204087-boracay-coconut-producer-agrarian-reform#cxrecs_s (accessed 20 December 2018).

Sen, D. (2014) 'Security guards accused of beating', *Phnom Penh Post*, 2 July. Available at: www.phnompenhpost.com/national/security-guards-accused-beating (accessed 13 September 2020).

Semple, K. (2016) 'Evictions by armed men rattle a Mexican tourist paradise', *New York Times*, 17 August. Available at: www.nytimes.com/2016/08/17/world/americas/mexico-tulum-corruption-evictions.html (accessed 4 June 2020).

Stabroek News (2018) 'In Cambodia, stalled Chinese casino resort embodies Silk Road secrecy, risks', 7 June. Available at: www.stabroeknews.com/2018/news/world/06/07/in-cambodia-stalled-chinese-casino-resort-embodies-silk-road-secrecy-risks/ (accessed 16 January 2019).

The Daily Guardian (2017) 'Town council mulls more special economic zones in Boracay', 13 March. Available at: https://thedailyguardian.net/local-news/town-council-mulls-special-economic-zones-boracay/ (accessed 31 December 2018).

The Manila Times (2018) 'Nurturing a sustainable partnership', 17 October. Available at: www.manilatimes.net/nurturing-a-sustainable-partnership/452718/ (accessed 6 January 2019).

Wiedeman, R. (2019) 'Who killed Tulum?', *The New York Times Magazine*, 18 February. Available at: www.thecut.com/2019/02/who-killed-tulum.html (accessed 4 June 2020).

WWF Philippines (2017) 'Sea turtles find a haven at Hamilo Coast'. Available at: wwf.org.ph/resource-center/story-archives-2017/sea-turtles-find-a-haven-hamilo-coast/ (accessed 6 January 2019).

4 Corporate resort development, residential tourism and resource grabbing

Once state-led tourism development and tourism zoning – as described in the previous chapter – have prepared the way for mass tourism streams, control of the tourism sector is often grasped by domestic elites in the country's major urban centres or by transnational hotel chains. These tend to be more interested in short-term profit-making than in establishing close ties with local communities (Telfer and Sharpley, 2008). Individual and communal properties of local residents can become an easy target for corporate resort developers who take advantage of weak legal recognition of customary land rights and low environmental and social safeguards. Rapidly rising land prices in popular tourism destinations lead to pressure on local landowners to sell their properties to the highest bidder. Enclave-style resort tourism in prime beach-front areas tends to cut off local communities from access to natural resources that are vital for sustaining their livelihoods, such as near-shore fishing grounds, mangrove forests and freshwater resources. The increasing popularity of tropical tourism hotspots among lifestyle and retirement 'migrants' or residential tourists adds to the stream of short-term mass tourists and fuel further processes of dispossession, enclosure and displacement.

For most tourists staying in luxurious beachfront resorts in tropical destinations, the idea that these places may have caused dispossession and forced displacement seems remote. Few may be aware of the fact that resorts and other segments of the tourism industry are prioritised in terms of receiving scarce freshwater resources, while local communities may experience chronic water shortages. Yet small islands and coastal tropical settings are particularly prone to tourism-induced conflicts over land and other natural resources, as they face challenges of resource scarcity and limited carrying capacity. The following sub-sections examine various forms of resource grabbing and displacement induced by corporate resort and residential tourism development in Indonesia, Vanuatu, Costa Rica and Mauritius.

Resort tourism and resource grabbing in the Indonesian archipelago

Like many Southeast Asian nations, the Indonesian government has promoted tourism growth at any cost, leaving very little space for alternative visions. The aggressive pursuit of mass tourism has led to widespread land grabbing, dramatic rises of land prices and increased competition for natural resources (Cole, 2017; Rosenberg Colorni, 2018). Such processes are particularly prevalent in Bali and islands in Eastern Indonesia.

Land grabbing and bay reclamation for tourism in Bali

On the Indonesian island of Bali, the dominant tourism sector has considerably changed property relations and dispossessed local people in several waves (Fagertun, 2017). Tourists have been lured into the 'exotic paradise' of Bali since the 1920s and 1930s when the island was part of the Dutch East Indies. The Dutch colonisers were the first to discover the island's economic potential for tourism, and the island was further popularised in Europe by a number of foreign-authored books and the works of visual artists that presented the island as a modern-day 'Garden of Eden' (Vickers, 2012; Cabasset, Couteau and Picard, 2017). Due to the reorganisation of Balinese society under Dutch colonial rule, local elites were able to occupy powerful administrative positions and gained greater control over the island's resources, which can be described as a first wave of dispossession (Fagertun, 2017).

In 1969, Suharto's autocratic and repressive government opened up the island to international tourism with the backing of a Master Plan drafted with support from the World Bank and the United Nations Development Programme (Cole, 2012). Large-scale dispossessions and forced expropriations were instigated in the 1980s and 1990s by turning customary land into state property, which then could be allocated for ambitious tourism and infrastructure projects – such as Nusa Dua in Jimbaran Bay – in the name of 'national development' (Warren, 2009). This large-scale land conversion from small-scale farming to tourism megaprojects was the second wave of dispossession (Fagertun, 2017).

After the fall of Suharto and the introduction of economic and administrative reforms in the late 1990s, local governments in Indonesia were able to impose and collect their own taxes; in Bali, land taxes have been based on the market value of a piece of land rather than its actual use (Cole and Browne, 2015). This regulation made land taxes unaffordable for farmers in the vicinity of tourist hotspots, thus forcing many of them into distress 'sales', usually in the form of 99-year lease contracts that bar locals from accessing the land for at least three generations. Only few farmers have been able to resist offers from the tourism industry to lease their land (Rosenberg Colorni, 2018).

It is estimated that the small island has lost nearly 25 per cent of its agricultural land – including a large share of its iconic rice terraces (Figure 4.1) – over the past 25 years (Rosenberg Colorni, 2018). This third wave of dispossession is still ongoing (Figure 4.2) and new megaprojects have been planned in some of the few remaining undeveloped areas of the island (Box 4.1).

Figure 4.1 Bali's iconic rice terraces, protected under UNESCO's World Heritage programme

Figure 4.2 Large tourist resort next to rice fields in Ubud, Central Bali

Box 4.1 The controversial Benoa Bay Reclamation Project

Until recently, Benoa Bay was one of the last remaining undeveloped areas in Southern Bali. In 2011, the bay – which covers a total area of 2,000 hectares, including 1,375 hectares of mangroves, five rivers, twelve village communities, and approximately 150,000 residents – was declared a maritime conservation area by presidential decree. Yet, in 2014, then President Yudhoyono revoked the protected status of Benoa Bay shortly before leaving office, thus annulling his earlier presidential decree and turning the area into a development and exploitation zone. Bali's former governor authorised the Indonesian developer PT Tirta Wahana Bali Internasional (TWBI) to develop an area of 838 hectares in Benoa Bay on a 30-year concession, extendable by another 20 years. TWBI had plans to develop the bay into a multi-billion US$ mega-tourism project, consisting of 12 artificial islands modelled after Dubai's Palm Islands. The developer's website brands the project as "[a] sustainable development project that integrates the origins of Balinese culture & traditional customs in the development of islands of resorts, theme parks, community areas, residential clusters to become the new iconic tourist-destination in Bali" (www.nusabenoa.com). Yet, in contrast to the claims of the developer that the project will 'revitalise' the bay and enhance the environment of the area through its 'eco-sustainable development concept', several studies found that the reclamation of Benoa Bay would have profoundly detrimental impacts on its fragile ecology and the local culture. Opponents have raised concerns that the project would damage coral reefs and mangrove areas, accelerate sedimentation of the bay, exacerbate water scarcity and pollution, and trigger widespread flooding of large parts of southern Bali.

Shortly after the reclamation plans became publicly known, opposition formed under the acronym ForBALI – The Bali People's Forum to Reject the Reclamation of Benoa Bay. This association brought together environmental activists, village leaders, politicians, journalists, academics, students, and artists and has become widely known by its slogan *Bali Tolak Reklamasi* (Bali Rejects Reclamation). Between 2015 and 2019, the movement organised protest marches, public fora and concerts to voice their determination to resist the project. It also collated a map with 22 sacred Hindu sites which were deemed at risk of disappearing should the reclamation project have gone ahead.

The controversy surrounding the planned project took an unexpected turn when the location and building permit of TWBI expired in August 2018. Under Indonesian law, if a permit is not renewed after four years, it is automatically cancelled. The company never received government approval to extend its permit, as the Ministry of Environment had not approved the developer's environmental impact assessment. Following the expiry of the permit, local activists called on President Widodo to re-establish the bay's protected area status. In October 2019, the Ministry of Maritime Affairs and Fisheries designated around 1,200 hectares of the Benoa Bay as a conservation zone, thereby banning reclamation projects, while allowing religious and cultural activities to continue.

Sources: Cabasset, Couteau and Picard, 2017; Ardhana and Farhaeni, 2017; Benge and Neef, 2018; Suriyani, 2018; Gokkon 2019

The rapid transformation of Bali into Indonesia's prime tourism destination has placed ever greater pressure on land and water resources, leading to the rapid decline of traditional wet-rice agriculture and dwindling local control over resource governance. Eighty-five per cent of the tourism industry is owned by non-Balinese, leading to a lack of accountability, gross power imbalances and a situation in which external investors can exploit natural resources without being directly affected by the negative implications of doing so (Cole and Browne, 2015).

The island's long-standing water management crisis has also caused enormous environmental problems and social conflicts (Cole, 2012; see also Benge and Neef, 2018). The diversion of water from agricultural areas to tourism hubs has led to growing distributional inequity between the tourism industry and local farmers (IDEP, 2015). Coupled with the unregulated exploitation of groundwater, this has allowed the tourism industry to consume water at a rate faster than its ability to replenish (Cole, 2012). Further pressure on water resources comes from the increasing number of residential tourists, mainly western retirees, who have been attracted by the island's pleasant climate, relative security, warm hospitality and cheap living cost (Bell, 2017).

Privatising beaches and constraining women's access to resources in Nusa Tenggara Timur

Having stretched Bali's tourism sector beyond its socio-ecological boundaries, the Indonesian government has targeted numerous islands in its eastern provinces for further promotion of tourism as outlined in its Master Plan for Acceleration and Expansion of Indonesian Economic Development 2011–2025. Nusa Tenggara Timur (NTT) – one of the archipelago's poorest provinces – plays a key role in reaching Indonesia's goal of welcoming 20 million tourists to the country. Provincial, district and municipal governments are encouraged to develop their long-term planning programmes with tourism as the leading sector alongside agriculture.

Labuan Bajo – situated on the western end of NTT's Flores Island – is undergoing a particularly rapid transformation from a once quiet port and fishing town to a bustling tourism destination. The main drawcard to this town is the nearby Komodo National Park, which is the habitat of the Komodo dragon, the world's largest lizard (Figure 4.3). The area also harbours some of the finest beaches and diving spots in Southeast Asia. The government has set a target of 500,000 visitors for the town by the year 2020, which is a fivefold increase compared to 2016 (Remmer, 2017).

The town is primarily inhabited by migrants from other Indonesian islands, and its population has increased from 10,000 in 2007 to 52,000 in 2014, with 80 per cent still living below the Indonesian poverty line (Cole, 2017). Due to the recent tourism boom, a large proportion of land in Labuan Bajo has been purchased by outside investors – from other parts of Indonesia or overseas – who control 70 per cent of the local tourism business (Remmer, 2017). The small islands around Komodo National Park (Figure 4.4) have been controlled by individuals and corporations that are part of national and international tourism business networks, some of which are foreign-owned businesses. Hence, the lion's share of tourism revenues is leaking to areas outside of Flores Island. On a positive note, large-scale resort developments by transnational hotel consortia have not occurred yet.

In 2003, Manggarai Barat became an autonomous regency by Indonesian Law No. 8/2003, with Labuan Bajo as its capital, thereby obtaining the authority to manage its own affairs, including the governance of the natural resources available in the territory (Cole, 2017). Local governments can now privatise formerly communal and public areas which triggered a major local conflict around public beach access in the Mabar community of Labuan Bajo (Box 4.2). Aside from increased incidences of land grabbing and rapidly rising

Figure 4.3 One of the iconic Komodo dragons in Komodo National Park

Figure 4.4 Small islands in Komodo National Park

property prices and rents, stakeholders in Labuan Bajo are also concerned with dwindling freshwater resources. Tourism development is competing for water supplies with residential water users and farmers, pushing up prices and violating human rights (Cole, 2017).

Box 4.2 Conflict between tourism investor and local community around public beach access in Labuan Bajo

In 2014, the former Provincial Governor of Nusa Tenggara Barat issued a permit to build a large hotel on a 4.2-hectare site by Pede Beach, the last remaining public beach in Labuan Bajo that local residents could access freely. The beneficiary of the 25-year land concession was a company owned by then Speaker of Parliament, Setya Novanto. The deal was later confirmed by the local government but ignited massive protests by local communities, student networks and church groups (the regency has a sizeable Christian population) from 2016 onwards. Despite sustained local resistance, construction of the hotel started in March 2017.

In May 2017, the Manggarai Student Alliance (AMANG) filed a complaint with the Corruption Eradication Commission against the Provincial Governor and the head of Manggarai Barat regency who allegedly issued the building permit amid financial irregularities. The allegations were that (1) the

Governor violated the Autonomy Law No. 8 of 2003 which required the NTT Provincial Government to hand over its existing assets in Western Manggarai to the Manggarai Barat regency when it was established as an autonomous region and (2) funds from the company were not sent to the provincial government's bank account but directly to the personal account of the Governor. Speaking to local media, the students stated that the report was part of an effort to crack down on corruption in Nusa Tenggara Timur, one of the most corrupt provinces in the country according to the Indonesia Corruption Watch survey. In the meantime, Setya Novanto who had been plagued by other allegations of corruption during his term started a 15-year jail term in April 2018.

Sources: Dauth, 2017; *UCANews*, 2017; UCAN India, 2017

Households in Labuan Bajo have been confronted with unregulated and unreliable water sources of doubtful quality for a long time (Remmer, 2017). Highly seasonal rainfall, irregular flows from local springs and saltwater intrusion into household wells have been common problems (Cole, 2017). Lack of regulation leads to uncontrolled and unmonitored withdrawal of groundwater (Remmer, 2017). Water is supplied to the piped water system from two local rivers, but this is prioritised to the hotel industry as the costs for pumping and purifying can only be recovered by charging business water rates – less than 25 per cent of the population have access to piped water (Cole, 2017). On average, the accommodation sector in Labuan Bajo consumes 275 litres of water per day and bed (Remmer, 2017), which is completely unsustainable, particularly when considering future tourism targets for the town. To make matters worse, the chronically inconsistent piped water supply in town has caused hotel owners to supplement it with tank water from trucks – affecting the cost and supply of this source for the community (Cole, 2017).

The cost for obtaining water from private sources is exceptionally high and places an enormous financial burden on Labuan Bajo's households (Remmer, 2017). Since Labuan Bajo is a highly patriarchal society, procuring household water is women's work. Women from poor households have to spend a large amount of time and effort on collecting and carrying water from public water supply stations, thereby diminishing their opportunity to engage in remunerative work (Cole, 2017). In sum, Labuan Bajo's water shortage problem has been gravely intensified by tourism development and has a disproportional impact on women.

Another region where women's livelihoods and access to resources have been negatively affected is on Rote Ndao, a regency in NTT province, situated south-west of the western tip of West Timor. Bo'a Beach at the western side of Rote Island has been the venue of an international surf competition for several years and has recently attracted a number of investors from other parts of Indonesia to raise the touristic profile of the island. In 2013/2014 access to the famous surf beach was blocked by the hotel development project of a company owned by the grandson of a former Indonesian president. This led to the relocation of the surf competition to another beach, Nemberala. Subsequently, new hotel developments reduced access to the beach for the local community living around Nemberala Beach which earned a decent income from near-shore seaweed production (pers. comm., D.V. Sinlae). Women in particular were involved in this profitable local business.

Several hotels breach Indonesian Law No. 1/2014 supported by President Regulation No. 51/2016, which prohibits construction within 100 meters from the highest tide point to the land. This law is designed to protect coastal ecosystems and communities from natural hazards and to provide space for public access to the beach, however, at high tide, the waves hit the fences built by hotels on the beach (pers. comm., D.V. Sinlae). The local community has vowed to resist the development of further beachside hotels and homestays, have torn down fences erected around hotel construction sites and even threatened to take violent action against foreign tourists.

The communal land on which the hotel premises at Bo'a beach have been built was previously transferred by the local community to the local government, based on the understanding that it would not be sold to private corporations. In breach of the agreement, the local government privatised the land by selling it to the tourism developer. The local authority of Rote Ndao argues that it is difficult to rectify this because the land ownership transfer occurred at a time when Rote Ndao was still part of Kupang Regency. Rote Ndao became an autonomous regency in 2002 under Indonesian Law No. 9/2002 (pers. comm., D.V. Sinlae).

The cases described above demonstrate how the decentralisation of tourism management and land regulations has actually aggravated infringement on the customary rights of local communities. The cases also show the particularly adverse impact of tourism-related land and resource grabs on women.

Tourism enclaves and proliferation of land leases in the South Pacific – the case of Vanuatu

Vanuatu is an archipelago of more than 80 islands located in the Southwest Pacific, about 2,300km northeast of Sydney, Australia. Its population of about 300,000 inhabitants is divided into more than 100 distinct linguistic and cultural groups. During colonial times, Vanuatu was known as the New Hebrides and subject to a rather unique Anglo-French colonial rule established in 1906. Throughout much of the 20th century, the indigenous ni-Vanuatu people were dispossessed of a great share of their customary land by both British and French settlers and missionaries who also introduced competing sets of laws and legal institutions (Farran and Corrin, 2017). Independence from the so-called 'condominium government' was only achieved in 1980, after demands for restitution of land alienated by the colonial powers for plantations, farms, settlements and churches could no longer be

suppressed (Farran, 2010). Since gaining independence, Vanuatu has benefitted from democratic rule, and its economy has seen relatively steady
growth rates, primarily due to a substantial rise in revenues from tourism
(Méheux and Parker, 2006).

The tourism sector in Vanuatu is characterised by a 'dualistic' structure,
where about one third of the foreign visitors arrive by air and stay in hotels,
resorts and guesthouses for an average of 8–9 days, while two thirds of visitors
arrive by cruise ship and stay for only one day without the need for accommodation in the country. Cruise tourists are primarily targeted by local tour
and cultural show operators, who are mostly Indigenous Ni-Vanuatu whose
small businesses are protected by the so-called 'Reserved Investments' clause
under the Foreign Investment Promotion Act. On the major islands, the hotel
industry – which is much more capital-intensive than tour operations – is
dominated by foreigners, who benefit from favourable investment conditions,
such as tax exemptions and relatively low lease rates for beachfront properties
(MTICNB, 2013). Most tourists stay on Efate Island where the capital Port
Vila is located (Figures 4.5 and 4.6).

The high demand for beachfront accommodation has led to a proliferation of land speculation among foreign investors. Land conflicts are
increasingly common, particularly in the rural areas of Efate, where customary land ownership is often ambiguous (see Box 4.3). Hierarchical
structures and differential access to land are predominant in communities,

Figure 4.5 Beach resort in the Pango district of Vanuatu's capital Port Vila

Figure 4.6 Cruise ship docking at Port Vila Harbour

and revenues from the proliferation of land leases benefit only a few. Land leases are overwhelmingly the providence of the chiefs; in Northern Efate, for instance, 80 per cent of the leases that have been signed off by individuals list a local chief as the lessor (McDonnell, 2015). Hence, only a small minority of the local population can actually take advantage of the booming lease market, while many community members feel the negative impacts of the continuing alienation of customary land in the form of leases to foreign investors.

Box 4.3 Customary land tenure system in Vanuatu: Strengthening Indigenous ownership or enabling land control by foreigners?

In precolonial times, land on the various islands in what is today's Vanuatu was acquired by simple occupation and building of the first meeting house. Ownership was established through physical evidence, such as graves, boundaries or planted trees, and through oral evidence. Intergenerational transfer of land was matrilineal in some areas and patrilineal in others.

Under British and French colonial rule (from 1906–1980) indigenous land was allocated to settler plantations, churches and public/administrative purposes. The concept of freehold and leasehold was introduced during that time, and about two thirds of land were in the hands of foreigners at some point. The 1980 Constitution restored indigenous land ownership across the newly independent country and provided that the rules of custom should form the basis for ownership, control and use of the land. Yet it was not always easy to identify the legitimate customary owners, leadership claims were often disputed and the number of counterclaimants was high, particularly in areas that had been most impacted by colonial settlement. Chiefly leaders often play the triple role of being customary landholders, figures of authority, and adjudicators of disputes.

In the early years after independence land leasing activity in Vanuatu was rather modest, confined primarily to agricultural leases of 30 or 40 years. Yet, with the advent of tourism and the associated diversification of the economy, non-agricultural leases with a longer duration (up to 75 years) were introduced. In 2013, the Vanuatu government introduced a new piece of legislation – the Custom Land Management Act – which was aimed at further strengthening customary land tenure and making it more difficult to alienate land through leases and sub-leases to foreign investors. However, the implementation of the Act has been constrained by a phase of political instability and Tropical Cyclone Pam in 2015.

Sources: Farran, 2008, 2010; Wittersheim, 2011; McDonnell, 2015;
Farran and Corrin, 2017

A villager in southern Efate reported in a focus group interview with the author that the lion's share of the benefits from a tourism land lease went to two families in the community, while the Council of Chiefs also received some part of the lease money and the actual customary landowners received nothing. A luxury hotel manager in northern Efate stated in an interview that many local people had leased out their land in the past without considering the long-term consequences. He further suggested that the benefits were not spread evenly within the community but were primarily controlled by the main chief in the village. In some communities, the position of the chief is contested, so part of the lease money would go to the lawyers.

Many Ni-Vanuatu landowners face problems borrowing financial capital off their customary land, which makes it difficult for them to start their own tourism business (MTICNB, 2013). Paradoxically, under a leasehold agreement, land can be subdivided, developed and even used as collateral for mortgage finance (Farran, 2008). By contrast, banks and other lending institutions in Vanuatu are reluctant to accept customary land as collateral for loans for various reasons: (1) if the borrower is not able to repay the loan, the lender cannot liquidate the asset as customary land cannot be sold; (2) if the lender leases the land to a third party (e.g. a foreign investor), this may cause social tensions in the community. Hence, land leasing to foreign investors is the main mechanism by which local landowners can generate monetary wealth, although the money obtained rarely reflects the real economic value of the land (McDonnell, 2015), let alone its less tangible but often more important cultural value.

According to Vanuatu's Department of Land, 80 per cent of the coastal land on Efate Island has been leased (Wittersheim, 2011), primarily to Australians, New Zealanders and – more recently – New Caledonians. Land is not only leased for resort development but increasingly also for building second homes for residential tourists. Most of this land has been fenced off or wired up (Figure 4.7). The enclosure of beachfront properties and – in some cases – entire islands (Figure 4.8) has substantially reduced access of local people to the sea.

> Today, we have to ask permission from the white men to access the sea, just like before independence.
>
> (Chief Roy Iasul, quoted in Wittersheim, 2011, p. 326)

Enclosure of coastal areas by tourism enclaves particularly affects women's livelihoods, as they engage mostly in fishing from the shore and on the reefs, while off-shore fishing is dominated by men (Government of Vanuatu, 2015). Furthermore, some of the local men are still able to negotiate access to the beach as they often know the hotel guards who hail from their own community.

Communities have complained about insufficient economic support from tourist resorts (particularly in the aftermath of disasters), the lack of transparency on lease agreements that were concluded several decades ago, and hotels

Figure 4.7 Fenced-off expat property development in Southern Efate

Figure 4.8 Rules of exception on an island under foreign lease

taking a major share of the profit from local village tours. There was also a sentiment that the benefits from both the leases and the tour operations were not spread evenly among the villagers.

Some of the resorts have lease agreements that include clauses stipulating that the land-owning community should be given priority in hotel staff recruitments, but the majority do not have such obligations. Large hotels, in particular, were also quick to lay off their Ni-Vanuatu staff after the 2015 Tropical Cyclone Pam which wreaked havoc on Efate Island and led to major damages to local communities and tourist resorts (Neef with Wasi, 2018). While many resorts offered short-term assistance to communities after the disaster, such as providing food, water and tools, only very few hotel businesses provided support for the long-term recovery in the communities from which they had leased land. One of the hotel managers interviewed in our 2016/2017 post-disaster study suggested that, while some villagers have improved their housing infrastructure, they are now confronted with issues of land scarcity, as many have leased out their land to tourism investors or expats involved in the tourism sector a long time ago (Neef with Wasi, 2018).

Although customary land is strongly protected by the country's legal framework and cannot be sold, Vanuatu has experienced a massive boom in the real estate market through long-term leases and sub-leases, primarily for resort development but more recently also for residential tourism projects. Reduced access to marine resources and coastal land is limiting the economic opportunities for most Ni-Vanuatu who still depend on farming and fishing for their subsistence. The volatility of employment opportunities in the tourism sector has been demonstrated in the aftermath of Tropical Cyclone Pam, when many Ni-Vanuatu employees in the tourism and hospitality sector were laid off (Neef with Wasi, 2018). Yet these post-cyclone impacts have been dwarfed by the 2020 COVID-19 pandemic that brought the tourism sector in Vanuatu to a complete standstill.

Residential tourism and transnational land investment in Central America and the Southern Indian Ocean

Residential tourism in developing countries, sometimes referred to as 'lifestyle migration' or 'retirement migration', is on the rise globally. In most cases, relatively wealthy citizens from Europe or North America – but more recently also from emerging economies, such as South Africa or Brazil – move temporarily or permanently to tourist destinations in developing countries where they buy or rent property and enjoy lower cost of living, better weather, beautiful scenery and a more relaxed lifestyle. Residential tourism implies the explicit granting of land concessions and tenure rights to foreigners. Both the number of residential tourists and the size of the related land investments keeps increasing in various countries in Asia, Africa, Latin America and the Caribbean. There is also growing competition among the destination countries to ease access and conditions for residential tourists (Aledo, 2008; Van Noorloos, 2014; Bell, 2017). In Panama and Honduras, for example, the governments have eliminated laws that once protected coastal lands from foreign ownership (Mollett, 2015).

Land alienation through residential tourism in Costa Rica

Partly as a consequence of the political instability and violence in the region in the 1980s, tourism developed in Central America more slowly than in the nearby regions of the Caribbean and southern Mexico (Cañada, 2010). Costa Rica was a notable exception with its relatively stable government and close economic ties with the US. Residential tourism investment in Guanacaste and Puntarenas provinces on Costa Rica's Pacific coast increased dramatically following 2003, when the volume of real estate investment overtook tourism expenditures (Cañada, 2010). Guanacaste province has seen several phases of historical land grabs, including (1) the Spanish conquest in the 16th century that wiped out the indigenous Chorotega civilisation, (2) transnational land acquisitions for cattle farming by North Americans in the late 19th century, which led to massive deforestation and instigated the introduction of private property rights and (3) large-scale land acquisitions for rice and cotton production by Costa Rican elites and US citizens in the 1960s and 1970s, which injected a speculative element into the local real estate market (Van Noorloos, 2014).

Following a major depression in the 1980s that affected the agricultural sector, the Costa Rican government invested heavily in improving its tourism infrastructure and tried to lure North American retirees into the country with generous tax incentives, which fuelled land speculation by foreign investors (Van Noorloos, 2011). Two thirds of the investments in the tourism industry are partly or fully financed by investors from the US and Canada, while collaborations between North American and Coast Rican investors are also common (Van Noorloos, 2014).

There is significant overlap between tourism, investment, and residency: short-term tourists often end up buying property and thereby become residential tourists on renewed visas; property owners rent out their property to short-term tourists, which turns many residential complexes into de facto hotels; shared ownership of properties is also common; and some investors buy properties without a single purpose, i.e. as holiday home, retirement property or simply as speculative asset (Van Noorloos, 2014). A number of transnational hotel chains, such as Hilton, Marriott and Sol Meliá have built a range of four- and five-star beach resorts with integrated leisure infrastructure to attract affluent tourists some of whom have turned into customers for second homes in the master-planned communities (Janoschka, 2009).

While the booming residential tourism industry has been a boon for the local real estate sector and speculative overseas investors, it is increasingly crowding out local small-scale tourism businesses, increasing low-paid employment and leading to greater inequalities. The global financial crisis of 2007/08 has also exposed the vulnerability of a sector that is highly dependent on North American economies and financial markets (Van Noorloos, 2014). In the post-crisis period, poverty and unemployment rates increased at a higher rate in Guanacaste province than in other regions; at the last national census of 2011, the poverty rates in the residential tourism hubs Guanacaste and Puntarenas were recorded at about 30 per cent, compared to the national average of 21.7 per cent (INEC, 2011).

For decades, Costa Rica has benefitted from strong and protective state institutions and laws, including a robust environmental legal framework. However, the implementation and control of environmental and spatial regulations have become more deficient through investor-led tourism development in Guanacaste (Janoschka, 2009). Lack of political will and inadequate human and financial capacities have triggered a host of socio-environmental problems, such as compromised conservation policies and coastal zone privatisation (see Box 4.4) as well as excessive water exploitation and small-scale displacement (Van Noorloos, 2014).

Box 4.4 Land tenure and the undermining of coastal regulations in Costa Rica

In contrast to most other countries in Central America, the majority of land in Costa Rica is private property. However, the law on the maritime-terrestrial zone (Law No. 6043, Ley sobre la Zona Marítimo Terrestre (ZMT), of 1977) stipulates rules for the use and protection of the first 200 metres of coastal land: the first 50 metres is inalienable public land, and the remaining 150 metres is designated as a restricted zone owned by government property. In this zone, land concessions can be issued (5–20 years renewable), while construction is allowed only under strict conditions. The ZMT law is meant to guarantee that coastal land is used for public benefit, that the socially and environmentally vulnerable coastal areas are protected, and that tourism is developed in a sustainable way.

However, reality shows a different picture: government regulations on the use of coastal land are not adhered to. A real estate market for coastal land has appeared; concessions are granted to foreign tourism companies and combining concessions increasingly leads to land concentration. Privatisation of the public inalienable zone (the first 50 metres) due to a complicated historical land tenure situation or entry barriers to public beaches is another growing problem. Coastal communities with land use permits are claiming more secure land rights, but these have so far been denied.

Source: Van Noorloos, 2019, p. 4

While the agricultural sector remains the industry with the highest water consumption, the water demand of the resort and residential tourism industry is inadequately recorded. La Voz de Guanacaste, an online service for local residents and international visitors, maintains that the wells registered for human use in the province have a capacity to supply nearly nine times more than the daily water needed for every resident of the province (García and Segnini, 2014). Yet, in some places, there have been reports of increasing water scarcity and fierce competition between the tourism industry and the residential sector over water use. For example, in the district of Sardinal, the Playa Panamá aquifer has been salinised as a result of overexploitation by hotels, which left the local communities without water supply (Cañada, 2018). As Van Noorloos (2017) found, the struggle over water use and resource conservation in Guanacaste is highly gendered, with women playing leading roles in local resistance movements against the extractivist practices of residential tourism. Women and underaged girls have also been impacted by a rapid surge of sex tourism in the province (Mowforth, Charlton and Munt, 2008; Van Noorloos, 2017). While the majority of Guanacaste residents still appear to have a largely positive view of tourism, this sentiment could rapidly change if the unbridled expansion of the resort and residential tourism sector continues.

The booming residential tourism industry and rising squatter communities in Mauritius

The Republic of Mauritius is located in the southern Indian Ocean 800km to the east of Madagascar and belongs geographically to the African continent. As a small, subtropical island, it has a land area of nearly 2,000km^2 and is surrounded and protected by coral reefs. With a population of close to 1.27 million (625 persons per km^2), it is one of the most densely populated countries in the world.

The island nation was under successive Dutch, French and British colonial rule which had a profound impact on postcolonial land rights. Mauritius has only two types of land tenure systems: one is freehold (private) land and the other is leasehold (government) land which may be leased upon application to the Ministry of Housing and Land Development (Olima, 2010). Land has always been a very sensitive issue in Mauritius, since the largest share of the land is owned by the descendants of slave owners (mostly Franco-Mauritian), who until recently were running most of the sugarcane farms and processing factories in the country (UN Habitat, 2012).

With the country's appeal to high-end tourists, Mauritius has made a quick transition from an economy based on a single agricultural commodity (sugar cane) to a diversified and booming tourism industry, catering to both short-term tourists and − more recently − to the rising residential tourist market (Wortman, Donaldson and van Westen, 2016). A focal point of residential tourism development is the Black River District, one of the country's poorest districts. Because of its scenic beauty, the district has attracted keen interest from real estate promoters looking for prime sites for the development of enclave resorts and gated residential tourist communities. The region has experienced one of the highest increases in property development projects in Mauritius with the number of housing units rising by 43 per cent from 2010 to 2012 (UN Habitat, 2012). Over the same period of time, foreign direct investment into real estate more than doubled across the small island nation, according to the Bank of Mauritius (2019). France, the UK and South Africa accounted for the largest contingents of foreign investors. A local real estate company boasts that the district capital Tamarin, formerly a sleepy fishing village, was searched on their website more than 485,000 times in 2017 compared to 356,000 in 2016 − a rise of 40 per cent (Lexpressproperty, 2018).

Besides upmarket hotels, several Integrated Resort Scheme projects (Box 4.5) have been developed in the Black River District. This recent development, combined with the multiplication of new residential areas for the Mauritian upper class, has profoundly changed the social and economic structure of the region. This is raising concerns among district councillors, particularly with regard to rising land prices and increased numbers of squatter settlements (UN Habitat, 2012).

Box 4.5 The Integrated Resort Scheme (IRS) in Mauritius

The Integrated Resort Scheme (IRS) is an initiative of the Government of Mauritius and the sugar industry, overseen by the Mauritian Board of

Investment, which provides non-citizens with the opportunity to purchase residential or resort-based property on the island. Prior to 2002, foreigners were not permitted to purchase property in Mauritius. The IRS promotes the construction of exclusive resorts and spacious gated communities comprising luxury, fully serviced properties along with a variety of high-quality amenities and facilities, such as marinas, golf courses, and wellness centres.

As per the IRS guidelines, the residential properties are sold for a minimum price of US$500,000 plus a fixed land registry charge of US$70,000, which comes with a right to residency in Mauritius for as long as they foreigner owns the property. A particular feature of the IRS is that resort developers are required to conduct a social needs analysis and a social impacts assessment, and divert a specified amount of money (around US $6,600) towards approved social projects in a one-off payment. Yet, there have been no studies to examine whether these social projects have had a sustainable impact on communities.

Source: Sharpley and Naidoo, 2010

A study by Wortman, Donaldson and van Westen (2016) identified the following areas of concern in relation to the residential tourism boom in Mauritius:

1 *Increased property prices and subsequent displacement*: since the rapid development of the Integrated Resort Scheme and new private commercial property projects in the district, the prices of properties in the region have skyrocketed. This has led to a steady growth in the number of squatter settlements has grown steadily, as many locals and migrant workers are not able to afford regular accommodation. A field survey by UN Habitat (2012) found that the Squatting Unit of the Ministry of Housing and Lands planned to relocate squatters, evicting those considered illegal without provision of alternative housing. District councillors find it increasingly difficult to identify land for social housing and recreational projects for citizens.

2 *Alienation caused by overdevelopment*: coastal and rural land is rapidly transforming into constructions sites for residential tourism development. Local residents have expressed fear of overdevelopment and overcrowding, as huge shopping centres replace local markets and gated communities crowd out traditional clusters of family homes. While many Mauritians remain largely positive about the increased investment flows, better infrastructure and more facilities through residential tourism, the sustainability of the current tourism development model seems questionable (Wortman, Donaldson and van Westen, 2012).

3   *Alienation due to the foreignisation of space*: the feeling that 'foreigners were taking over the island' was expressed by many respondents in Wortman et al.'s study. The fact that most foreign property buyers are Caucasian causes frustration among impoverished locals and conjures up feelings of exclusion and foreign domination in a country with a long history of slavery and oppression by colonisers. Some respondents mentioned how they were offended by the way gated communities were advertised as being conveniently fenced off against the locals, intensifying the feeling of 'us versus them' (Wortman, Donaldson and van Westen, 2016).

It seems paradoxical and cynical that the European Commission is supporting upgrade programmes for mushrooming squatter settlements in one of Africa's richest countries, while EU citizens continue to roam Mauritius' real estate market website to secure their own slice of paradise. And the irony does not stop there: Mauritius has not only become a haven for residential tourists but also a tax haven for foreign investors grabbing land elsewhere (Pearce, 2012).

Concluding remarks

The cases of Indonesia and Vanuatu are a stark reminder that formalised customary land tenure does not necessarily provide a strong defence against transnational land deals; in fact, these legal frameworks can become enablers of land grabbing when farmers are forced into distress sales or leases (as in the case of Bali) and when customary land owners cannot use their land as collateral but can only valorise it in monetary terms by putting it on the foreign-dominated long-term lease market (as in the case of Vanuatu). Residential tourism development and transnational land investments in Costa Rica and Mauritius – incentivised by these countries' investor-friendly governments – have induced a rapid 'foreignisation of space' (cf. Zoomers, 2010) where autochthonous residents feel like aliens in their own country. In all four cases, natural resources, such as freshwater, near-shore fisheries and mangroves that have played a pivotal role in local livelihoods, become increasingly enclosed by rapid resort development for temporary visitors and residential tourists (cf. Gössling et al., 2012; Tourism Concern, 2012; Becken, 2014; LaVanchy, 2017). The impacts are often gendered, with women particularly affected. Some communities in Bali, Indonesia and in Guanacaste, Costa Rica have resisted these developments, yet with mixed success rates.

The next chapter will examine various cases of opportunistic disaster capitalism in the aftermath of major disaster events, whereby tourism developers and local governments often collude to turn 'crisis into opportunity', make use of the 'blank slate' left behind by the disaster and force disaster-affected communities off their land. The case studies are chosen from Honduras, Thailand, Haiti and the Philippines.

References

Aledo, A. (2008) 'De la tierra al suelo: La transformación del paisaje y el Nuevo Turismo Residencial [From the land to ground: The transformation of the landscape and the new residential tourism]', *Arbor: Ciencia, Pensamiento y Cultura*, 184 (729): 99–113.

Ardhana, I.P.G. and Farhaeni, M. (2017) 'The study of the impact for social culture toward the planning of reclamation for Benoa Bay in Bali', *AIP Conference Proceedings 1844*. doi:10.1063/1.4983437.

Becken, S. (2014) 'Water equity – Contrasting tourism water use with that of the local community', *Water Resources and Industry*, 7–8: 9–22.

Bell, C. (2017) '"We feel like the King and Queen": Western retirees in Bali, Indonesia', *Asian Journal of Social Science*, 45 (3): 271–293.

Benge, L. and Neef, A. (2018) 'Tourism in Bali at the interface of resource conflicts, water crisis, and security threats' in Neef, A. and Grayman, J.H. (eds) *The Tourism-Disaster-Conflict Nexus*. Emerald Publishing: Bingley, pp. 33–52.

Cabasset, C., Couteau, J. and Picard, M. (2017) 'La poldérisation de la baie de Benoa à Bali: vers un nouveau puputan? [Benoa Bay reclamation in Bali: Toward a new puputan?]' *Archipel* 93: 151–197.

Cañada, E. (2010) *Tourism in Central America, Social Conflict in a New Setting*. Fundación PRISMA/Alba Sud Research Paper. Available at: www.albasud.org/publ/docs/32. en.pdf (accessed 8 January 2019).

Cañada, E. (2018) *The Struggle for Water: Reducing the Spread of Tourism in Costa Rica*. *Tourism Watch*, Newsletter No. 90 (February 2018). Available at: www.tourism-wa tch.de/en/content/struggle-water (accessed 20 December 2018).

Cole, S. (2012) 'A political ecology of water equity and tourism: a case study from Bali', *Annals of Tourism Research*, 39 (2): 1221–1241.

Cole, S. (2017) 'Water worries: An intersectional feminist political ecology of tourism and water in Labuan Bajo, Indonesia', *Annals of Tourism Research*, 67: 14–24.

Cole, S. and Browne, M. (2015) 'Tourism and water inequality in Bali: a social-ecological systems analysis', *Human Ecology*, 43: 439–450.

Fagertun, A. (2017) 'Waves of dispossession: The conversion of land and labor in Bali's recent history', *Social Analysis*, 61 (3): 108–125.

Farran, S. (2008) 'Fragmenting land and the laws that govern it', *Journal of Legal Pluralism*, 58: 93–113.

Farran, S. (2010) 'Law, land, development and narrative: A case-study from the South Pacific', *International Journal of Law in Context*, 6 (1): 1–21.

Farran, S. and Corrin, J. (2017) 'Developing legislation to formalise customary land management: Deep legal pluralism or a shallow veneer?', *Law and Development Review*, 10 (1): 1–27.

Gössling, S., Peeters, P., Hall, M.C., Ceron, J., Dubois, G. and Lehmann, L. (2012) 'Tourism and water use: Supply, demand, and security. An international review', *Tourism Management*, 33: 1–15.

Government of Vanuatu (2015) *Post-Disaster Needs Assessment – Tropical Cyclone Pam, March 2015*. Government of Vanuatu: Port Vila, Vanuatu.

INEC (2011) *Censo Nacional de Población y Vivienda* [National Population and Housing Census] – *2011*. Instituto Nacional de Estadísticas y Censos [National Institute of Statistics and Censuses]: Costa Rica.

Janoschka, M. (2009) 'The contested spaces of lifestyle mobilities: regime analysis as a tool to study political claims in Latin American retirement destinations', *Die Erde* 140: 251–274.

LaVanchy, G.T. (2017) 'When wells run dry: Water and tourism in Nicaragua', *Annals of Tourism Research* 64: 37–50.

McDonnell, S. (2015) '"The land will eat you": Land and sorcery in North Efate, Vanuatu' in Forsyth, M. and Eves, R. (eds) *Talking it Through: Responses to Sorcery and Witchcraft Beliefs and Practices in Melanesia*. Australian National University Press: Canberra, pp. 137–160.

Méheux, K. and Parker, E. (2006) 'Tourist sector perceptions of natural hazards in Vanuatu and the implications for a small island developing state', *Tourism Management*, 27 (1): 69–85.

Mollett, S. (2015) 'The power to plunder: Rethinking land grabbing in Latin America', *Antipode* 48 (2): 412–432.

Mowforth, M., Charlton, C. and Munt, I. (2008) *Tourism and Responsibility: Perspectives from Latin America and the Caribbean*. Routledge: London, New York.

MTICNB (2013) *Vanuatu Strategic Tourism Action Plan 2014–2018*. Ministry of Tourism, Industry, Commerce & Ni-Vanuatu Business: Port Vila, Vanuatu.

Neef, A. with Wasi, S.A. (2018) *Disaster Response and Recovery of the Tourism Sector: The Case of Vanuatu in the Aftermath of 2015 Cyclone Pam*. Unpublished Research Report, Auckland, New Zealand.

Olima, W.H.A. (2010) *Property Taxation in Anglophone East Africa: Case Study of Mauritius*. Working Paper, Lincoln Institute of Land Policy, University of Pretoria: South Africa.

Pearce, F. (2012) *The Land Grabbers: The New Fight over who Owns the Earth*. Bacon Press: Boston, MA.

Remmer, S. (2017) *Tourism Impacts in Labuan Bajo*. Swisscontact WISATA: Denpasar.

Richter, L. (2008) 'Tourism policy-making in Southeast Asia: A twenty-first century perspective' in Hitchcock, M., King, V.T. and Parnwell, M. (eds) *Tourism in Southeast Asia: Challenges and New Directions*. Northern Institute of Asian Studies Press: Copenhagen, pp. 132–145.

Rosenberg Colorni, R. (2018) *Tourism and Land Grabbing in Bali: A Research Brief*. Transnational Institute: Amsterdam.

Sharpley, R. and Naidoo, P. (2010) 'Tourism and poverty reduction: The case of Mauritius', *Tourism and Hospitality Planning & Development*, 7 (2): 145–162.

Telfer, D.J. and Sharpley, R. (2008) *Tourism and Development in the Developing World*. Routledge: London, New York.

Tourism Concern (2012) *Water Equity in Tourism – A Human Right, A Global Responsibility*. Available at: www.tourismconcern.org.uk/wp-content/uploads/2014/10/Water-Equity-Tourism-Report-TC.pdf (accessed 22 January 2019).

UN Habitat (2012) *Mauritius: Black River Urban Profile*. United Nations Human Settlements Programme: Nairobi.

Van Noorloos, F. (2011) 'Residential tourism causing land privatization and alienation: New pressures on Costa Rica's coasts', *Development*, 54 (1), 85–90.

Van Noorloos, F. (2014) 'Transnational land investment in Costa Rica: Tracing residential tourism and its implications for development' in Kaag, M. and Zoomers, A. (eds) *The Global Land Grab: Beyond the Hype*. Fernwood Publishing: Halifax & Zed Books: London, New York, pp. 86–99.

Van Noorloos, F. (2017) 'A women's world or the return of men? The gendered impacts of residential tourism in Costa Rica' in Archambault, C. and Zoomers, A. (eds) *Global Trends in Land Tenure Reforms: Gender Impacts*. Routledge: London, New York, pp. 78–94.

Van Noorloos, F. (2019) 'Tourism turning real estate: How to deal with residential tourism investment in the global South?', LANDac Policy Brief 1. Utrecht University: Utrecht. Available at: https://usercontent.one/wp/www.landgovernance.org/wp-content/up loads/2019/07/LANDac_Policy_Brief_01.pdf (accessed 19 January 2019).

Vickers, A. (2012) *Bali: A Paradise Created.* Tuttle Publishing: Singapore.

Warren, C. (2009) 'Off the market? Elusive links in community-based sustainable development initiatives in Bali' in Warren, C. and McCarthy, J.F. (eds) *Community, Environment and Local Governance in Indonesia: Locating the Commonweal.* Routledge: London, New York, pp. 197–226.

Wittersheim, E. (2011) 'Paradise for sale. The sweet illusions of economic growth in Vanuatu', *Journal de la Société des Océanistes,* 133: 323–332.

Wortman, T., Donaldson, R. and van Westen, G. (2016) '"They are stealing my island": Residents' opinions on foreign investment in the residential tourism industry in Tamarin, Mauritius', *Singapore Journal of Tropical Geography,* 37: 139–157.

Zoomers, A. (2010) 'Globalisation and the foreignisation of space: seven processes driving the current global land grab', *The Journal of Peasant Studies,* 37 (2): 429–447.

Media sources / websites

Bank of Mauritius (2019) 'Foreign Direct Investment Tables', Available at: www.bom. mu/foreign-direct-investment-tables (accessed 7 January 2019).

Dauth, M. (2017) 'AMANG reported Lebu Raya, Gusti Dula, CEO of PT SIM and Setya Novanto to KPK', 17 May. Available at: www.melanesiahotnews.com/hukum/17/ 05/2017/1394/amang-reported-lebu-raya-gusti-dula-ceo-of-pt-sim-and-setya-novan to-to-kpk/ (accessed 10 January 2019).

García, E. and Segnini, G. (2014) 'Guanacaste produces 8.7 times more water than amount needed for human use', *La Voz de Guanacaste,* 7 April. Available at: https:// vozdeguanacaste.com/en/guanacaste-produces-8-7-times-more-water-than-amount-needed-for-human-use/ (accessed 6 January 2019).

Gokkon, B. (2019) 'Bali mangrove bay is now a conservation zone, nixing reclamation plan'. Available at: https://news.mongabay.com/2019/10/bali-benoa-bay-mangro ves-conservation-reclamation/ (accessed 10 July 2020).

IDEP (2015) 'Bali water protection program'. Available at: www.idepfoundation.org/ bwp (accessed 14 February 2017).

Lexpressproperty (2018) 'Residential real estate: the 5 most popular regions in Maur-itius (and why they are so attractive …)'. Available at: www.lexpressproperty.com/ en/news-advices/accommodation/residential-real-estate-the-5-most-popular-regions -in-mauritius-and-why-they.html (accessed 10 January 2019).

Suriyani, L.D. (2018) 'As Bali reclamation project dies, activists seek conservation status', *Mongabay,* 30 August. Available at: https://news.mongabay.com/2018/08/as-bali-rec lamation-project-dies-activists-seek-conservation-status/ (accessed 20 January 2019).

UCAN India (2017) 'Indonesia students report governor for corruption', 22 May. Available at: http://m.ucanindia.in/news/indonesia-students-report-governor-for-co rruption-34830.html (accessed 10 January 2019).

UCANews (2017) 'Church accuses Indonesia of riding roughshod over poor', 3 April. Available at: www.ucanews.com/news/church-accuses-indonesia-of-riding-rough shod-over-poor-/78814 (accessed 10 January 2019).

5 Tourism expansion, land grabbing and resistance in post-disaster contexts

Many studies have focused on the positive role of tourism following major disaster events. It is commonly believed that tourism has the potential to revitalise disaster-stricken communities and contribute positively to the recovery process (e.g., Ritchie, 2009; Marshall, 2015; Van Strien, 2018). Yet, in many cases, rapid and unfettered tourism development in the aftermath of a catastrophic event may allow rogue investors to prey on the plight of disaster-affected people and turn the disaster into an economic opportunity for their own benefit (Cohen, 2011; Neef et al., 2018; Neef and Grayman, 2018).

Klein (2007a) coined the term 'disaster capitalism' (cf. Box 5.1), making reference to the reshaping of the post-disaster landscape by private investors – often in conjunction with governments, donors and financial institutions – while masking their real economic agenda with a discourse of rehabilitation and 'building back better'. Disaster capitalism – particularly when exercised by the tourism industry – has the potential to ignite a host of conflicts over post-disaster recovery processes and access to natural resources.

Box 5.1 Disaster capitalism

Schuller (2008, p. 20) defines 'disaster capitalism' as "[n]ational and trans-national governmental institutions' instrumental use of catastrophe ... to promote and empower a range of neoliberal capitalist interests." It refers to the use of disasters and acute crises as opportunities for the private and public sectors to capitalise on temporary or permanent vulnerabilities and to push for policies and practices that are likely to be rejected in times of social and moral order (Klein, 2007b; Forgie, 2014; Schuller and Maldonado, 2016; Pyles, Svistova and Ahn, 2017).

One of the first reported cases of disaster capitalism occurred in the wake of Hurricane Mitch which lashed the Central American countries of Honduras, Guatemala and Nicaragua in October 1998 and killed more than 11,000 people. The first section of this chapter discusses how the Honduran government opened the country to foreign tourism investors in the immediate aftermath of the disaster.

Disaster capitalism reached another level following the 2004 Indian Ocean Tsunami, which claimed the lives of more than 227,000 people in 14 countries and displaced around 2.5 million people throughout the region. Following this unprecedented disaster event, the Sri Lankan government imposed a no-building zone of 200 meters from the high-water mark in the eastern and northern parts of the island country, but exempted resorts from the buffer zone regulation (Klein, 2007b). This measure prevented about 30 per cent of the tsunami-affected population from returning to their land, while opening ample opportunities for investors in the tourism industry (Cohen, 2011). Governments in other tsunami-affected countries, such as Indonesia and India also declared arbitrary setback limits with no consideration of local topography and without consulting local governments and planning authorities, while exempting hotels from the ban on building directly at the shoreline (Mulligan and Shaw, 2007; Mowforth and Munt, 2016). This led to allegations that policy making had been captured by elite interests and that the true purpose of the setback limits was the relocation of poor fishing families and squatter communities from prime beach locations to make way for tourism development.

Thailand is another country that was heavily impacted by the 2004 Indian Ocean Earthquake and Tsunami and where disputes over coastlines ensued in the aftermath. A set of three case studies in this chapter discusses conflicts over access to land and other natural resources that affected Indigenous communities along the kingdom's Andaman coast, facing the Indian Ocean. The third section in this chapter explores disaster capitalism following the 2010 earthquake in the Caribbean island nation Haiti, where tourism has been proposed as a key recovery strategy. The final case study examines a major corporate land grab that occurred following 2013 super typhoon Haiyan on Sicogon Island in the Philippines.

Post-disaster livelihood displacement among Garifuna communities, Honduras

In 1998, Hurricane Mitch brought utter devastation to several Central American countries. In Honduras, an estimated 7,000 people died, some 12,000 persons remained missing (presumed dead), and about two million people (more than one third of the small country's population) were left homeless (Stonich, 2000). At the time of the catastrophic event, tourism in Honduras had already become one of the most vital and continuously growing sectors of the country's economy. The sector was, at the time, concentrated around the Bay Islands and the cultural heritage site of Copan, which remained relatively unscathed from the disaster. In the midst of the recovery process, the Honduran government engaged a US-based public relations firm to develop a tourism campaign strategy together with the Honduran Institute of Tourism (Stonich, 2000). The strategy even included a brand of disaster tourism under the name 'The Trujillo Project', whereby US American tourists were lured with discounted airfares into vacationing on a tropical beach, with removal of debris, planting of trees and restoring of turtle nesting sites all part of the holiday package (Mowforth and Munt, 2016).

Only a few months after the disaster, the Honduran congress passed several laws to facilitate the privatisation of airports, seaports, highways and parts of the water sector (Klein, 2007b). Further extraordinary judicial measures opened up opportunities for the tourism sector, most importantly the reform of Article 107 of the Honduran Constitution to legalise the purchase of land and ownership of infrastructure by foreign investors on the Caribbean and Pacific coasts and the country's islands (Stonich, 2000). Provisions under the 1999 Tourism Incentive Law waived the income tax over a period of 15 years for new tourism establishments and exempted them from paying taxes and any other duties on goods and equipment imports during their construction and start-up operations (Loperena, 2017).

Development of the tourism sector was explicitly prioritised in the *Master Plan for National Reconstruction and Transformation* (with its Spanish acronym PMRTN) published in 1999. 'Ecotourism' was presented as the most viable strategy of both boosting tourist numbers and preserving pristine nature in this poverty-stricken country. A military coup in 2009 caused a temporary decline of tourism numbers, but in 2011 the newly elected president announced in a high-profile international conference that 'Honduras is Open for Business' and presented tourism as one of six key investment areas (Loperena, 2017). Major emphasis was now on large-scale tourism developments, and Tela Bay with its six Garifuna communities was selected as one of the focal sites. Back in 1998, the then Minister of Tourism had already identified the North Coast as a "tourist developer's dream" with its "600 kilometers of uninhabited beach" (quoted in Mollett, 2014, p. 37). Conjuring up the tourism imaginary of an empty beachfront area waiting to be valorised by 'ecotourism' resort development is a prototypical example of the 'idle land' discourse described in Chapter 2. In addition to enclosure by ecotourism development, the declaration of both terrestrial and marine protected areas along the North Coast has contributed to limiting access of Garifuna communities to their customary subsistence farming areas and fishing grounds (Mollett, 2014; Loperena, 2016). These protected areas include the Cayos Cochinos Marine Protected Area and the Jeannette Kawas National Park (Brondo, 2013; Loperena, 2016). Miranda (2009) estimated that about 60 per cent of the Garifuna communities lived within the confines of protected areas.

Several tourist and real estate megaprojects have been planned within the buffer zones of Tela Bay's Jeannette Kawas National Park since the early 2000s (Loperena, 2016). The financial infrastructure for tourism megaprojects in Honduras is composed of capital from national elites, foreign investors and loans from the Central American Bank for Economic Integration and the Inter-American Development Bank (Loperena, 2017). The World Bank also got involved through its 'Honduran Poverty Reduction Strategy', approved in 2001, which funded a four-year Sustainable Coastal Tourism Project that was aimed at improving environmental planning capacity and tourism destination management along the municipalities of the North Coast (Brondo, 2013). The Honduran state has played its part through enacting further pro-tourism legislation, implementing neoliberal economic policies, creating special economic development zones, and providing the state-security apparatus in the form of its police and military forces (Loperena, 2016). Through a special provision, the state has legalised foreign ownership in places like Tela Bay by declaring coastal land a 'tourism priority' and classifying such land as 'urban', thereby opening the real estate market to foreigners (Brondo, 2013; Mollett, 2014).

The liberalisation of the land market in combination with the expansion of protected areas has put enormous pressure on the Garifuna who fear the dismantlement of their ancestral, communally held territories (cf. Box 5.2). Despite being predominantly subsistence fishing and farming communities, the six Garifuna communities have been classified as 'urban' which exposed them to market forces (Mollett, 2014). Giving in to mounting pressure by the state and corporate investors, some Garifuna have sold parcels of communal land since the early 2000s out of fear of losing the land to the tourism sector without compensation (Thorne, 2004). Violent interventions by park rangers and naval officers – including the 2008 killing of a Garifuna fisherman accused of fishing illegally in a protected area – further instilled fear and anxiety among Garifuna communities which have been labelled as environmental threats due to their alleged refusal to abide by marine protection measures (Brondo, 2013; Loperena, 2016). Ironically, tourists and private investors are hailed as stewards of the environment by the Honduran Institute of Tourism and the National Institute of Conservation and Forestry Development (Loperena, 2016).

**Box 5.2 Garifuna communities on the North Coast of Honduras:
A brief ethno-legal history**

The Garifuna people are descendants of African slaves and two indigenous groups originally from South America – the Arawak and the Carib Indians. In 1797, the British colonial power deported 5,000 Garifuna, also known as Black Caribs, from the Caribbean island St. Vincent to another British colony Roatán, an island among the Islas de la Bahia archipelago about 65 kilometres off the North Coast of Honduras. From Roatán, the Garifuna people migrated to the Honduran mainland and across the Atlantic coast of Central America, including Nicaragua, Guatemala and Belize. In Honduras, the Garifuna established communities along the North Coast of Honduras in the early 19th century where they engaged in subsistence farming and fishing. Garifuna connections to coastal lands are grounded not only in their livelihoods but also in important cultural rituals, such as ancestor worship. Their distinct cultural traditions, including language, dance and music, were declared a Masterpiece of the Oral and Intangible Heritage of Humanity by UNESCO in 2001. Despite this global recognition of their culture, the

Garifuna continue to face discrimination and harassment within the Honduran socio-political and economic system.

Traditionally, the Garifuna have only known collective land ownership, often vested in women due to the group's matrilineal and matrilocal customs. Historical documents show that Garifuna communities in Tela Bay had been in possession of registered *ejidal* (community) land in the early 20th century. Female land ownership and collective titles have been weakened by a World Bank instigated land registration programme that focused on the distribution of private land titles in disregard of the Garifuna's legal traditions. The Garifuna have suffered from various waves of dispossession, starting with the expansion of US-owned banana plantations in the late 19th century, which forced many Garifuna to abandon residential areas and agricultural lands. More recently, the Garifuna have faced pressure from the expanding tourism industry and the delineation of new protected areas.

Source: Brondo, 2013; Mollett, 2014; Jubis, 2015; Loperena, 2016

One tourist mega-complex – inaugurated in 2013 – extends over 500 hectares along three kilometres of coastline and boasts an 18-hole golf course, several five-star hotels, an equestrian centre, 400 private villas, shopping centres, bars, restaurants, and pools (Mollett, 2014). The resort was funded through a public-private partnership between the Honduran Institute of Tourism and the Tela Bay Touristic Development Society whose president is also the head of the financial group FICOHSA – one of the largest financial institutions in Central America (Loperena, 2017). While the resort was not directly involved in the physical displacement of the Garifuna, investments in the area's infrastructure combined with land speculation led to a slow but steady erosion of Garifuna's territorial rights (Loperena, 2017), a process that Mollett (2014, p. 40) has coined "displacement-in-place".

Prior to the establishment of the resort, neighbouring Garifuna communities possessed full ownership titles to their ancestral territories, although not all were fully recognised by the government. Research by the Council of Hemispheric Affairs found that the implementation of the 1992 Agrarian Modernization Law – backed by the US administration and the World Bank – led to the expansion of Tela's city boundaries and stimulated transactions of ancestral lands without consent of the Garifuna communities (Jubis, 2015). In a gross misuse of its power, the municipality of Tela sold a portion of the Garifuna ancestral territories to a local corporation well beyond the real market value of the land and later issued construction permits for the development of several large-scale tourism projects (Brondo, 2013; Jubis, 2015).

Yet this form of state violence has been countered since the late 2000s by strong grassroots resistance led by the Ethnic Community Development Organization (*Organización de Desarrollo Etnico Comunitario* – ODECO) and the

Black Fraternal Organisation of Honduras (*Organización Fraternal Negra Hon-dureña* – OFRANEH) which organised a series of local protests and took the case to the Inter-American Commission on Human Rights (IACHR) (Brondo, 2013). In the midst of massive local protests and ongoing territorial disputes between the investors and the Garifuna communities, Honduran national police and military forces tried to violently remove the Garifuna from their lands in 2014 (Loperena, 2017). In 2015, the IACHR ruled that the Honduran government was to restore ancestral land to the Garifuna commu-nity and compensate them for their losses (Agudelo, 2019). Yet, five years later, the ruling has yet to be enforced.

To date, the conflicts between tourism development, militarised conservation and Indigenous rights to ancestral lands remain largely unresolved, and Garifuna communities continue to live in a constant state of fear. Since September 2019, at least five Garifuna leaders and land rights defenders have been killed, according to the United Nations High Commissioner for Human Rights (OHCHR) in Honduras, and in July 2020, five Garifuna community leaders were kidnapped at gunpoint by armed men in police uniforms (Lakhani, 2020). At the time of writing this chapter, their fate remains unknown.

Post-tsunami dispossession of Indigenous seafaring people in Phang Nga and Phuket, Southern Thailand

Thailand's Andaman coast is home to three distinctive yet inter-related Indi-genous communities known collectively as *chao leh*, or sea people. The Moken, Moklen and Urak Lawoi who comprise the *chao leh* communities have a combined population of around 7,000 in Thailand. Until recently, they were seafaring communities with cultural identities and subsistence practices rooted in marine and coastal resources. Nearly all *chao leh* communities in Thailand were severely impacted by the 2004 Indian Ocean Earthquake and Tsunami which killed several thousand local residents and foreign tourists in the pro-vinces Phang Nga, Phuket and Krabi. In the aftermath of the tsunami, the *chao leh* were confronted with a new brand of disaster capitalism, orchestrated by powerful public and private actors. The following subsections are based on the author's long-term research into the struggle of the *chao leh* against disposses-sion and displacement (Neef et al., 2018). Box 5.3 presents a brief overview of the three Indigenous communities.

Box 5.3 Indigenous communities of the Andaman Coast, southern Thailand, in the aftermath of the 2004 Indian Ocean Tsunami

Moken

The Moken, the most 'mobile' of the *chao leh* communities, have used the islands of Koh Surin as their temporary settlements between sea travels for centuries. Their seasonal migration around Koh Surin had them classed as stateless and thus kept under the watchful eye of authorities especially once the islands collectively became a national park in 1981. Although the tsunami destroyed their entire settlement, the Moken were able to draw on traditional knowledge which warned them of the imminent disaster and saved all but one member of their community.

Moklen

Unlike the Moken, both the Moklen and Urak Lawoi communities have longer histories of 'semi-permanent settlement' along the coast. In the case of the Moklen, one of their settlements Baan Tungwa suffered catastrophic impact including the loss of lives and survivors' livelihoods. With all of their subsistence means destroyed, the Moklen were temporarily relocated and had attempts made against them by the government to permanently relocate them further inland.

Urak Lawoi

Further south in Baan Rawai reside the Urak Lawoi. Among the three groups, Urak Lawoi communities have the longest history of living in 'semi-permanent settlement' along the coast. Given their location at the southern tip of Phuket Island, they were least impacted by the tsunami. Though their customary land rights have never been formally acknowledged in the region, the Urak Lawoi have a long history in Baan Rawai dating back to at least the late 19th century.

Source: Attanavich et al., 2015; Robinson and Drozdzewski, 2016;
Neef et al., 2018

The Moken and post-disaster tourism in the Surin Islands Marine National Park

The Moken's informal residential rights on Koh Surin have long been marked by the presence of cultural artefacts predating the conversion of the islands to a national park. Near their settlements and cemetery grounds the Moken would erect *lobong*, or spirit poles, to demarcate both cultural and geographical space and hold annual ceremonies to instil their spiritual significance (Figure 5.1). Although *lobong* were appropriated by the national park and commandeered as the park's symbol, the ceremonies were still respected and permitted by park authorities (Neef et al., 2018).

Following the tsunami, the relationship dynamic between the Moken and the park authorities changed dramatically. Post-tsunami recovery encouraged a new wave of tourism to Koh Surin, prompting authorities to tighten regulations over the Moken's settlements and drastically reduce their access to forest,

Figure 5.1 Moken spirit pole (*lobong*) on Koh Surin

coastal and marine resources. Under these new regulations the Moken have been legally confined to a singular location in northern Koh Surin and denied their customary migration practices. Having been demobilised by the authorities, the formerly semi-nomadic community began to face issues of increased resource use and slower resource regeneration, and a concentration of demand for employment. Post-tsunami relief attempted to address the latter issue by capitalising on the distinctive cultural element that the Moken had to offer tourism on Koh Surin. By commodifying the cultural practices and beliefs of the Moken, tourism development essentially turned the settlement into a human zoo, appropriating the profitable parts of their culture and heavily restricting their customary livelihood practices.

With strict legal regulations over resource use within the national park, employment opportunities have drastically decreased for the Moken. Prohibited from felling trees for building their traditional boats (*kabang*), collecting sea-snails and commercial fishing, the only options left for them are selling souvenirs (Figure 5.2) or working for the park – both forcing them into dependency on seasonal tourist streams and the appropriation of their culture in order to make a living.

The cramped living conditions in a single space under the Park authorities' scheme proved disastrous in February 2019, when a fire broke out in one house, quickly spreading through the village and destroying 61 out of the 81 thatched-roofed houses, leaving 273 people homeless and without belongings (*Bangkok Post*, 2019; Smillie, 2019). While Park authorities and the Thai

Figure 5.2 Moken villagers on Koh Surin selling souvenirs to tourists

military provided quick post-disaster recovery support, the Moken were not consulted in the rebuilding efforts and design of their new homes, despite their wealth of Indigenous knowledge in vernacular architecture (pers. comm., M. Attavanich). This disregard of Indigenous cultural practices may well expose the Moken to future disaster risks.

Land conflict resolution attempts with the Moklen in Baan Tungwa

The Moklen community of Baan Tungwa and their way of living fundamentally changed following the impact of the tsunami. A sizable share of their community including their religious leader was lost to the disaster alongside traditional housing, boats and fishing equipment, livestock, and gardens – all of which played significant roles in maintaining the social and cultural fabric of their community (Attavanich et al., 2015). The community also belonged to the 89 out of 428 villages affected by the tsunami that suffered from extremely uncertain land tenure status and were embroiled in land conflicts in the aftermath of the unprecedented disaster (ACHR, 2006).

Left without homes, the Moklen were temporarily relocated immediately following the tsunami. This relocation risked becoming permanent when local government tried to coerce them into re-settling further inland. Though this relocation was ostensibly for their safety, the reality was that the government had earmarked their traditional lands for public infrastructure development. Yet the entire community refused to relocate and eventually were assisted to rebuild their community in its original location (Attavanich et al., 2015; Neef et al., 2018). During the rebuilding process local officials incessantly intimidated the community in further attempts to make them relocate. This eventually culminated in a land-sharing agreement wherein the community were able to lease a majority share of the land, which was promised to later be turned into a permanent communal land title, while the remaining portion was allocated for public use. This agreement was hailed by academics and NGO representatives as a breakthrough in securing the residential rights of Indigenous seafaring people along the Andaman coast (ACHR, 2006). Yet the local government did not keep their promise of formally recognising the Moklen communal property rights, instead only extending the lease period (Neef et al., 2018).

Meanwhile, severe restrictions have been imposed over coastal and marine spaces that became reserved for tourist use only. New beachfront developments were authorised by local government (Figure 5.3). Coral reefs and mangrove areas that once provided their main sources of subsistence and income were declared off-limits to the Moklen. The nearby beachfronts are increasingly monitored in order to keep the 'unsightly' *chao leh* away from tourists. Interviews conducted by the author between 2013 and 2017 revealed how access to nearshore fishing grounds has increasingly been restricted. The only areas which Moklen have not been denied access to are situated far from the area that they formerly used (Figures 5.4 and 5.5). Customary burial grounds have also been razed by developers (Attavanich et al., 2015).

Figure 5.3 Change in access to foreshore resources

Figure 5.4 Beachfront development in the former livelihood area of Baan Tungwa

Figure 5.5 Moklen from Baan Tungwa fishing in the foreshore of the public beachfront area

Although the Moklen were able to continue residing on their ancestral lands, the perpetual undermining of their rights and traditional practices has had a dire impact on their community. With many forced into unemployment and facing severe socio-economic hardship, post-tsunami tourism recovery efforts continue to adversely affect the Moklen to this day.

Recognising customary land rights of the Urak Lawoi in Baan Rawai

The Urak Lawoi community presents a very distinctive case from the previous one in both the impact they sustained from the tsunami and their capacity to respond to tourism development in their territory. Despite not being strongly impacted by the 2004 Indian Ocean tsunami, the Urak Lawoi still became the victims of a series of land grabbing attempts made against them by both local elites and external investors. Yet, in contrast to most *chao leh* communities in Phang Nga, the Urak Lawoi in southern Phuket were able to exercise continued resistance against these land grabs.

The rights over their territory are largely divided between two major parties with the western area being targeted by three local Thai businessmen and -women, and the eastern part being claimed by a large tourism developer (Figure 5.6). Deprived of their customary ownership rights, the Urak Lawoi have been forced to concentrate their entire community of some 2,000 people across little more than 250 small houses in a very crammed space, resulting in extreme slum-like conditions.

Figure 5.6 Map of the disputed sites in Baan Rawai

Despite their already marginal living conditions, their situation was worsened when three local entrepreneurs began to assert ownership over the western part of their territory and sue villagers for refusing to leave the area. To this end, a total of 30 lawsuits have been filed which have impacted the lives of at least 121 villagers. Since 2017, however, this impact has not been as detrimental as originally feared on account of the following evidence that irrefutably proved the customary land rights of the Urak Lawoi over the territory:

- Human remains were discovered in an area excavated for a new house construction that were forensically identified by the Thai Department of Special Investigation as an ancestor of members of their community. The remains were accompanied by items that were used in traditional burial ceremonies, further supporting their claims of customary occupation.
- Old school records proved their attendance of local schools, and some ID cards formally recognised their occupation of the land since at least the late 19th century.

Given the insurmountable evidence, six of the lawsuits have already been dropped, and it is anticipated that the remainder will similarly be ruled in favour of the Urak Lawoi. Yet, unfortunately for the Urak Lawoi, the recognition of their customary land rights has not been as easily recognised in the eastern part of their territory with a Bangkok-based tourism developer, Baron World Company, claiming to have had legal rights to the lands since 1965. The land in question is a public land area that lies adjacent to their residential area and is intended to be developed into luxury beach villas through a total investment of about US$18.5 million.

The first claim the company made to the land was in 2015 when they illegally erected a seawall to bar villagers from accessing the beach and destroyed a local waterway, a village boat dock, and a spiritual shrine in the process. Although ordered by the Marine Department to be demolished, the order was undermined by the Deputy Governor of Phuket Province. Left without any support, villagers chose to remove the seawall themselves and rebuild their ceremonial shrine in its original location (Figure 5.7).

In 2016, the company had two violent confrontations with Urak Lawoi villagers (Figure 5.8) disputing this same area of land, which eventually forced state agencies to intervene and protect the rights of the Urak Lawoi. Mediation processes attempted to remedy the conflict by offering villagers a nearby site for their ceremonial shrine and activities instead. The offer was refused and compelled community representatives to deliver a petition to the Minister of Justice, urging him to investigate the land attacks against them. In response, the company filed a number of lawsuits against the community, most notably for illegal occupation of their territories, and for 31 million Thai Baht worth of damages. Every such case has been ruled in favour of the Urak Lawoi and upheld their rights to occupy their customary lands.

Figure 5.7 The shrine of Baan Rawai villagers

Figure 5.8 Boulders placed by company workers to block villagers' access to their ceremonial area

Similar to the case in the western part of their territory, the forensic findings, old student records, and community IDs, all supported their claim to have settled in this area decades prior to the company receiving the official land title from the government. What made their case particularly compelling, however, were photographs and video recordings documenting the late King Bhumibol Adulyadej visit to their well-established village in 1959. As the kingdom was still mourning the loss of their revered King, his former support of the community had a decisive impact on the court rulings.

The legal recognition that the Urak Lawoi have achieved is but one improvement to the range of socio-economic and political pressures that negatively impact their health and wellbeing. Without decent economic opportunities and healthy living conditions, their socio-economic and cultural situation remains extremely precarious.

Disaster capitalism and tourism development in post-earthquake Haiti

Haiti is situated between the North Atlantic and the Caribbean Sea, occupying the smaller western part of the island of Hispaniola which it shares with the Dominican Republic. Once deemed the most lucrative French colony, it became the first independent Caribbean state and the first Black republic in 1804 after a decade-long struggle against French colonisation and enslavement (Séraphin, 2018). Its subsequent development was hindered by illegitimate French reparation claims for former slaveholders and the US American occupation from 1915 to 1934 (Daut, 2020). Five years after the end of the US occupation, Haiti's Destination Management Organisation (DMO) was established under President Stenio Vincent (1939–1941) who identified tourism as a potential revenue source for the country. Ten years later, the Port-au-Prince International Exhibition of 1949 instigated a short-lived golden period for Haiti when it became the most popular tourist destination in the Caribbean for about a decade (Séraphin and Butcher, 2018).

Yet, the political turmoil of the Father and Son Duvalier era (1957–1986) along with a distorted image of Haiti being an epicentre of HIV/AIDS in the 1980s threw tourism in the former 'Pearl of the Antilles' into a deep crisis. Only a few international tourism operators made deals with the Haitian government under dictator Jean-Claude 'Baby Doc' Duvalier, including Club Med which opened a resort on government-leased land in 1981 (Séraphin, 2018) and the cruise ship company Royal Caribbean which obtained a concession for a 105 hectare (260 acre) peninsula to construct the enclave resort Labadee, named – rather ironically – after French slave owner and plantation baron Marquis de La'Badie (Walker, 2010). The resort is surrounded by a ten-foot (three-metre) fence, protected by heavily armed private security guards and only open to tourists descending on the resort from the company's cruise ships for a brief stopover (Weeden, 2015). Tourists are prohibited from leaving the resort, while only a small number of locals are allowed to enter the resort against a fee and sell souvenirs (Séraphin and Butcher, 2018). Royal Caribbean is not charged an annual payment for the land concession but pays a small fee of US$12 to the Haitian government for each of their cruise ship tourists. Apart from this dual-enclave tourism model (cruise ships as floating enclaves, Labadee as a land-based enclave) which reportedly has been the single largest contributor to the tourism sector in Haiti, it is alleged that only the presence of NGOs and other international agencies kept the tourism and hospitality sector afloat from the 1960s to 2010 (Séraphin, 2018).

On 12 January 2010, Haiti was struck by a powerful earthquake, killing between 220,000 and 300,000 people and displacing at least 1.5 million people. The densely populated region around the capital Port-au-Prince was among those most heavily affected, with most of the infrastructure damaged or destroyed. The following months saw the implementation of a so-called 'Action Plan for the Reconstruction and National Development of Haiti' which included the establishment of an Interim Haiti Recovery Commission (IHRC) and the implementation of hundreds of reconstruction projects driven by bilateral donors (primarily the United States, Canada, Brazil and France), international financial institutions and a myriad of foreign investors (Dupuy, 2010; Pyles, Svistova and Ahn, 2017). The IHRC – which had originally been conceived by the US State Department and ratified by the Haitian Parliament – was co-chaired by then Haitian Prime Minister Jean-Max Bellerive and former US President Bill Clinton who had been appointed UN Special Envoy for Haiti about eight months prior to the disaster (Dupuy, 2010). This earlier appointment by then UN Secretary-General Ban Ki-moon followed the devastation caused by a series of tropical storms and hurricanes, with the former US President expected to restore donor confidence and promote foreign private investments in the troubled country (Charles, 2020).

By effectively surrendering the country's sovereignty to the IHRC for a period of 18 months, Haiti continued its long history of foreign interference in its internal affairs since gaining independence in 1804 (Dupuy, 2010). In the year following the disaster, the US government reportedly awarded over 1,500 contracts for reconstruction projects of which only about 20 went to Haitian businesses (Panchang, Bell and Field, 2012). Luxury tourism – alongside mining, mango production and an expanded sweatshop industry – was regarded as a quick-fix solution for rebuilding the Haitian economy (Forgie, 2014). Encouraged by Bill Clinton, Irish billionaire Denis O'Brien, the CEO of Digicel – one of the largest mobile phone networks in the Caribbean – built a US$45 million Marriott hotel in the capital Port-au-Prince (completed in 2015) and contributed US$12 million to the restoration of the century-old Iron Market, a major tourist hotspot, that had been destroyed by the earthquake (Charles, 2020). Both investments have proved largely unsuccessful: the government-owned, uninsured Iron Market was partially destroyed by a fire in early 2018, and the Marriott hotel struggled with the continuing civil unrest and international travel advisories that discourage vacationing in Haiti.

Bill Clinton and then US Secretary of State Hillary Clinton backed a new post-earthquake candidate for the Haitian presidency, popular singer Michel Martelly, who ran his campaign on the slogan 'Haiti is open for business' (Schuller and Tsu, 2014). Following his inauguration in 2011, he embarked on a particularly aggressive tourism development strategy. In May 2013, he issued a decree that declared Île à Vâche, a 45.5km^2 pristine island off the southwestern coast of Haiti, a 'zone of public utility', thereby revoking all private property rights (Ives, 2014). A follow-up decree prohibited any further constructions on the island without prior authorisation (Schuller and Tsu, 2014). Martelly's vision for the US$250 million project was to build numerous hotels, villas and bungalows, access roads, an 18-hole golf course and other tourist attractions (Adams, 2014; Jeannite and Lapointe, 2016). Potential investors were lured by the promise of a 15-year tax holiday and infrastructure developments, such as the construction of a US$13 million international airport, renovations of the public market and improved drinking water supply (Kushner, 2015). The government was aided by the lack of a comprehensive land tenure system (Kelly, Deaton and Amegashie, 2019) and the Haitian Constitution which declares all coastal land – along with springs, rivers, water courses, mines and quarries – as part of the State's public domain.

Despite this, many of the island's 14,000 inhabitants who practice subsistence farming and fishing had been rightful owners of their land, a rare occurrence in a country where only five per cent of the total territory has been mapped by the National Cadastral Office (Kelly, 2019). Ironically, it was the country's brutal dictator François 'Papa Doc' Duvalier (1957–1971) who had distributed 2,500 acres (~1,000 hectares) of land to the local population after expropriating a senator who had turned the island into his private fiefdom (Chery, 2014). When a Dominican company started to bulldoze local farmers' land and coconut trees for an access road in 2013 and ground was broken for an international airport in the same year, the islanders organised themselves under several protest groups, such as Citizen Action for Île à Vâche (ACI) and the Organisation of Île à Vâche Farmers (*Konbit Peyizan Ilavach*, KOPI) (Schuller and Tsu, 2014; Chery, 2014). The protests took various forms, including peaceful demonstrations, road blockades and the closing of schools and businesses (Ives, 2014).

To squash the local protest movement, the government deployed more than 120 heavily armed police officers and soldiers to the island (Chery, 2014). At the height of the conflict, a popular local protest leader was jailed without charges and a proper trial (Adams, 2014; Ives, 2014). According to KOPI, hundreds of islanders lost parts of their farmland and access to communal resources to the construction of the airport runway over the course of 2014 (Chery, 2014; Kushner, 2015). Intervention attempts by the Minister of Tourism and Minister of Extreme Poverty failed. The protesters refused to engage in negotiations unless their demands to lift the president's public-utility decree, to remove the armed forces from the island and to free the protest leader would be met (Adams, 2014).

After a standoff of several months, the Haitian government decided to call off the tourism development project on Île à Vâche in 2015, as potential investors had become increasingly concerned about the ongoing protests and the opacity of the local land tenure situation (Kushner, 2015). Yet irreversible damage had already been done through the destruction of local farmers' coconut plantations and the enclosure of communal land by infrastructure development, such as the – now abandoned – airstrip and access road. Hence, it is likely that the socially and economically exclusionary enclave model represented by Royal Caribbean's Labadee resort on Haiti's north coast will remain the principal form of tourism in the country despite the controversies and risks surrounding it (Séraphin, 2018). In 2016, one of Royal Caribbean's flagship cruise ships, the Freedom of the Seas, was met by local protesters in small boats and had to turn away from Labadee (Golden, 2016). The suspension of cruising during the global COVID-19 pandemic has left the resort idle for the most part of 2020.

Corporate land grabs on Sicogon Island in the Wake of Typhoon Haiyan (Philippines)

In November 2013, super typhoon Haiyan (known in the Philippines as 'Yolanda') struck around 117 coastal cities and municipalities in 14 provinces and six regions across the Philippines. The typhoon – one of the strongest ever recorded – killed at least 6,300 people, made more than 900,000 families homeless and displaced about 205,000 families from their land. It had a devastating impact on the survivors' livelihoods and the coastal environment, where coral reefs and mangroves were destroyed (Calvan, 2015).

After the typhoon, then President Benigno Aquino III declared a no-build zone of 40 meters from the coast, including where people used to live. This triggered confusion and anger among local governments, civil society groups, and communities affected by typhoon Haiyan who wanted to return to their land (Manahan, 2017). However, the government was able to draw on Section 108 of the Philippine Fisheries Code of 1998 which mandates the Department of Agriculture (DA) through the Bureau of Fisheries and Aquatic Resources (BFAR) to provide 'safe and secure settlements' for fishing families near their fishing grounds. The government projected that nearly 25,000ha of land were needed to accommodate displaced families and that the costs of resettlement would amount to more than US$1.7 billion (Calvan, 2015).

Since the reconstruction costs were deemed too high for the national government to shoulder them alone, it entered into a unique public-private partnership to support recovery efforts in the aftermath of the typhoon. For the first time in the history of the Philippines, it asked 20 corporations to spearhead post-disaster rehabilitation, giving them the opportunity to choose which islands they wanted to 'adopt' and rebuild (Ambrose and Majeed, 2018). Ayala Corporation, one of the oldest and largest business conglomerates in the Philippines with a market capitalisation of US$11.5 billion (as of June 2018), strategically chose the island of Sicogon. The clean slate left by the disaster in combination with government-imposed no–dwelling zones and legal provisions for relocation of communities to 'safe zones' provided a convenient opportunity for the company to grab formerly occupied land under the pretext of providing relief efforts and improving human security (Uson, 2017).

The small island of Sicogon is located in the central province of Iloilo and home to about 1,100 families who engage in artisanal fishing and backyard farming. A local elite family holds the ownership rights to 70 per cent of the island under the corporate acronym SIDECO and decided to engage in a strategic partnership with Ayala Corporation (Uson, 2017). The extremely uneven land tenure system on the island is typical for many parts of the country, where the concentration of land ownership in the hands of a few influential families (known as the 'landed elite') has its roots in the colonial regimes of the Spanish and Americans (cf. Chapter 3). The remaining 30 per cent of the island is made up of public and forest lands held by the state (Uson, 2015).

About 20 years before typhoon Haiyan ravaged the island, 335 hectares of land (approximately 40 per cent of the landowning family's land) had been earmarked for land distribution to about 250 local families under the Comprehensive Agrarian Reform Program (CARP) (Manahan, Cruz and Carranza, 2015; Uson, 2017). CARP is one of two major 'pro-poor' land policy programmes in the Philippines, alongside the Indigenous Peoples Rights Act (IPRA) (see Box 3.2). The planned land redistribution under CARP was strongly opposed by the landowning family which used its connections with state officials and local business councils and referred to the alleged 'unproductivity of the agricultural sector' to make a case for excluding the island from CARP (Uson, 2015). Yet, despite the landowners' resistance and strong support from the then Minister of Tourism, the regional director of the Department of Agrarian Reform (DAR) issued a Notice of Coverage in July 2009, confirming that the 335ha would be placed under CARP (Conserva, 2014; Uson, 2017).

The Ayala-SIDECO alliance used the desperate situation of the Sicogon residents to rid the island of its inhabitants and free its shores for their tourism development plans. A week after the typhoon, Ayala-SIDECO offered two options to the disaster-affected people: the first option was a PHP 150,000 (about US$3,200) cash payout if they agreed to leave the island and move to a destination of their own choice; the second option was free housing, water and electricity provided by the corporate alliance on the mainland (Manahan, 2017). Only if they accepted either of the two options, they would be allowed to receive disaster relief goods (Uson, 2017).

In return for accepting one of the two offers, the 'beneficiaries' had to sign a far-reaching agreement that contained the following:

1   they acknowledge that the family firm is the registered owner of the 809 hectares;
2   they confirm that they have no right to stay on Sicogon island;
3   they waive their rights to their land and all their cases against the landowner;
4   they promise to destroy their house and permanently leave the island;
5   they acknowledge that the market payoffs are the landowner's humanitarian support to their family;
6   they withdraw their CARP application; and
7   they confirm that they signed on their own volition and would not reclaim ownership of its property on the island.

(Uson, 2017, p. 423)

The majority of the Sicogon residents – most of whom were not eligible to receive land titles under CARP – accepted the offer out of desperation. Those who decided to stay and fight on faced harassment and encroachment by the corporate alliance's security guards and anti-reform government actors (Uson, 2017). Additional pressure was exerted by the no-dwelling zone policy imposed by the government, as discussed earlier. Having no other alternative, more than 200 remaining Sicogon families decided to settle in a portion of a 282-hectare public forest land area.

After a one-year impasse, the Federation of Sicogon Island Farmers and Fisherfolk Association (FESIFFA) entered into a compromise agreement with Ayala-SIDECO in November 2014, with strong support from civil society groups. The Compromise Framework Agreement (CFA) provided that Ayala-SIDECO would prioritise local residents in employment for work in the resort and provide 30 hectares of land for residential area, 40 hectares of land for conventional farming, a livelihood support and capacity building fund of about US$846,000 and a US$1.7 million land development fund (Laiko, 2018).

However, none of the land reforms that were agreed upon in the CFA were subsequently delivered, which motivated FESIFFA to resume its advocacy work for their land rights. In April 2017, resistance leaders went to the capital and organised a camp-out in front of the Department of Environment and Natural Resources (Manahan, 2017). Meanwhile, in August 2017, Ayala Hotels and Resorts Corporation – a subsidiary of Ayala Corporation – entered into a memorandum of agreement with SIDECO and an individual to develop Sicogon Island into a new leisure destination, buying land on the island at the price of US$6.2 million from SIDECO in September 2017 (Ayala Corporation, 2018). In the same month, the chairman and CEO of Ayala Corporation Jaime Augusto Zobel de Ayala was honoured by the United Nations as a 'pioneer of sustainable development'. At the occasion of receiving the award, he was quoted saying "I believe that now, more than ever, a deeper engagement with society is indispensable to the survival and success of private enterprises. From both a practical and moral standpoint, businesses cannot thrive in an environment rife with economic inequity" (Schnabel, 2017, n.p.).

Meanwhile, the economic and social inequities on Sicogon Island continued. In December 2018, around 40 Sicogon farmers and fisherfolks held another camp-out in front of the Department of Agrarian Reform (DAR) Central Office as part of their 'Reclaiming the land, water and livelihood rights of Yolanda survivors in Sicogon Island' campaign. They demanded the issuance of a Cease and Desist Order (CDO) to stop the on-going construction of tourist facilities in the area as well as the reinstatement of land coverage under CARP (Laiko, 2018). Eventually, in March 2019, the DAR ordered Ayala Corporation and private owners of high-end resorts on Sicogon Island to stop any development activities on the disputed parcel of land being claimed by FESIFFA, signifying a major victory for Sicogon farmers and fishers over corporate land grabbers (*Philippine Daily Inquirer*, 2019).

Concluding remarks

Tourism presents its most violent features in the aftermath of major disasters. Such events provide particularly fertile ground for tourism investors to take advantage of traumatised and incapacitated communities and to turn a temporary crisis into a protracted one. The 'crisis discourse' legitimises swift and decisive policy measures that would otherwise be strongly resisted. Since tourism is often afforded a pivotal role in disaster recovery, opportunistic and predatory actors have an easy job of dispossessing legitimate customary landholders. As the case studies from Honduras and Thailand have shown, practices of dispossession can be both material and discursive (cf. Loperena, 2016) and do not always involve the physical displacement of communities. 'Displacement-in-place' (Mollett, 2014, p. 40), whereby Indigenous people become part of the marketed tourism landscape, while their ancestral lands and natural resources are enclosed and encroached upon, is also common. Yet all four cases have also shown signs of hope. In the long run, predatory practices of disaster capitalists do not remain unchallenged, as evidenced by the partially successful counter-movements along the north coast of Honduras, on Thailand's Rawai Beach, on Haiti's Île à Vâche and on Sicogon Island in the Philippines.

As the next chapter will show, governments and corporate investors do not only capitalise on 'natural' disaster events. Conflicts and post-conflict situations also provide ample opportunities for a variety of state and non-state actors to seize land from oppressed Indigenous groups and ethnic minorities, former enemies and occupied populations for tourism purposes. This problematic entanglement of armed conflict, militarism and tourism will be discussed using the cases of Myanmar, Sri Lanka, Bangladesh and Israel/Palestine.

References

ACHR (2006) *Tsunami Update*. Available at: www.achr.net/upload/downloads/file_22122013021601.pdf (accessed 23 October 2020).

Agudelo, C.E. (2019) 'The Garífuna community of Triunfo de la Cruz versus the State of Honduras: Territory and the possibilities and limits of the Inter-American Court of human rights verdict', *Latin American and Caribbean Ethnic Studies*, 14 (3): 318–333.

Attavanich, M., Neef, A., Kobayashi, H. and Tachakitkachorn, T. (2015) 'Change of livelihoods and living conditions after the 2004 Indian Ocean Tsunami: The case of the post-disaster rehabilitation of the Moklen Community in Tungwa Village, Southern Thailand' in Shaw, R. (ed.) *Recovery from the Indian Ocean Tsunami: A Ten-Year Journey*. Springer Publishers: Heidelberg, New York, Dordrecht, London, pp. 471–486.

Brondo, K.V. (2013) *Land Grab: Green Neoliberalism, Gender, and Garífuna Resistance in Honduras*. University of Arizona Press: Tucson.

Calgaro, E. and Lloyd, K. (2008) 'Sun, sea, sand and tsunami: examining disaster vulnerability in the tourism community of Khao Lak, Thailand', *Singapore Journal of Tropical Geography*, 29: 288–306.

Calvan, D. (2015) *Land, Property and Tenurial Rights in a Changing Coastal Environment*. Land Watch Asia, Asian NGO Coalition for Agrarian Reform and Rural Development. Available at: www.angoc.org/wp-content/uploads/2015/12/Land-Property-etc_final.pdf (accessed 23 October 2020).

Cohen, E. (2011) 'Tourism and land grab in the aftermath of the Indian Ocean Tsunami', *Scandinavian Journal of Hospitality and Tourism*, 11 (3): 224–236.

Dupuy, A. (2010) 'Disaster capitalism to the rescue: The international community and Haiti after the Earthquake', *NACLA Report on the Americas*, 43 (4): 14–19.

Forgie, K. (2014) 'US imperialism and disaster capitalism in Haiti' in Forte, M.C. (ed.) *Good Intentions: Norms and Practices of Imperial Humanitarianism*. Alert Press: Montreal, pp. 57–75.

Jeannite, S. and Lapointe, D. (2016) 'La production de l'espace touristique de l'Île-à-Vache (Haïti): illustration du processus de développement géographique inégal [Production of tourism space on Île-à-Vache (Haiti): illustration of the process of unequal geographic development]', *Études Caribéennes* [Caribbean Studies], 33–34, April–August 2016. Available at: https://journals.openedition.org/etudescaribeennes/8810 (accessed 6 June 2020).

Kelly, L.D., Deaton, B.J. and Amegashie, J.A. (2019) 'The nature of property rights in Haiti: Mode of land acquisition, gender, and investment', *Journal of Economic Issues*, 53 (3): 726–747.

Klein N. (2007a) 'Disaster capitalism: The new economy of catastrophe', *Harper's Magazine* 315: 47–58.

Klein, N. (2007b) *The Shock Doctrine: The Rise of Disaster Capitalism*. Penguin Group: London.

Loperena, C.A. (2016) 'Conservation by racialized dispossession: The making of an eco-destination on Honduras's North Coast', *Geoforum*, 69: 184–193.

Loperena, C.A. (2017) 'Honduras is open for business: Extractivist tourism as sustainable development in the wake of disaster?', *Journal of Sustainable Tourism*, 25 (5): 618–633.

Manahan, M.A. (2017) 'Advancing justice after climate disaster in the Philippines', *Farming Matters*, June: 42–44.

Manahan, M.A., Cruz, J. and Carranza, D. (2015) 'Standing on contentious ground: Land grabbing, Philippine style', *Focus on the Global South – Policy Review*, 1 (6): 3–9.

Miranda, M. (2009) *Areas Protegidas y las Comunidades Garífunas* [Protected Areas and Garifuna Communities]. Organización Fraternal Negra Hondureña – OFRANEH: La Ceiba.

Mollett, S. (2014) 'A modern paradise: Garifuna land, labor, and displacement-in-place', *Latin American Perspectives*, 41 (6): 27–45.

Mowforth, M. and Munt, I. (2016) *Tourism and Sustainability: Development, Globalisation and New Tourism in the Third World* (4th edition). Routledge: London, New York.

Mulligan, M. and Shaw, J. (2007) 'What the world can learn from Sri Lanka's post-tsunami experiences', *International Journal for Asia Pacific Studies*, 3 (6): 65–91.

Neef, A. and Grayman, J.H. (2018) 'Introducing the tourism-disaster-conflict nexus' in Neef, A. and Grayman, J.H. (eds) *The Tourism-Disaster-Conflict Nexus*. Emerald Publishing: Bingley, pp. 1–31.

Neef, A., Attavanich, M., Kongpan, P. and Jongkraichak, M. (2018) 'Tsunami, tourism and threats to local livelihoods: The case of indigenous sea nomads in southern Thailand' nexus' in Neef, A. and Grayman, J.H. (eds) *The Tourism-Disaster-Conflict Nexus*. Emerald Publishing: Bingley, pp. 141–164.

Pyles, L., Svistova, J. and Ahn, S. (2017) 'Securitization, racial cleansing, and disaster capitalism: Neoliberal disaster governance in the US Gulf Coast and Haiti', *Critical Social Policy*, 37 (4): 582–603.

Ritchie, B.W. (2009) *Crisis and Disaster Management for Tourism*. Channel View Publications: Bristol, Buffalo, Toronto.

Robinson, D.F. and Drozdzewski, D. (2016) 'Hybrid identities: Juxtaposing multiple identities against the "authentic" Moken', *Identities: Global Studies in Culture and Power*, 23 (5): 536–554.

Schuller, M. (2008) 'Deconstructing the disaster after the disaster: conceptualizing disaster capitalism' in Gunewardena, N. and Schuller, M. (eds) *Capitalizing on Catastrophe: Neo-Liberal Strategies in Disaster Reconstruction*. Alta Mira Press: Walnut Creek, CA, pp. 17–27.

Schuller, M. and Maldonado, J.K. (2016) 'Disaster capitalism', *Annals of Anthropological Practice*, 40 (1): 61–71.

Séraphin, H. (2018) 'The past, present and future of Haiti as a post-colonial, post-conflict and post-disaster destination', *Journal of Tourism Futures*, 4 (3): 249–264.

Séraphin, H. and Butcher, J. (2018) 'Tourism management in the Caribbean: The case of Haiti', *Caribbean Quarterly*, 64 (2): 254–283.

Stonich, J.C. (2000) *The Other Side of Paradise: Tourism, Conservation, and Development in the Bay Islands*. Cognizant Communication Corporation: New York, Sydney, Tokyo.

Thorne, E.T. (2004) 'Land rights and Garífuna identity', *NACLA Report on the Americas*, 38 (2): 21–25.

Uson, M.A.M. (2015) *Grabbing the 'Clean Slate': The Politics of the Intersection of Land Grabbing, Disasters and Climate Change – Insights from a Local Philippine Community in the Aftermath of Super Typhoon Haiyan*. ISS Working Paper No. 603. International Institute of Social Studies of Erasmus University, Rotterdam. Available at: https://repub.eur.nl/pub/77539 (accessed 2 January 2019).

Uson, M.A.M. (2017) 'Natural disasters and land grabs: The politics of their intersection in the Philippines following super typhoon Haiyan', *Canadian Journal of Development Studies*, 38 (3): 414–430.

Van Strien, M. (2018) 'Tourism business response to multiple natural and human-induced stressors in Kathmandu, Nepal nexus' in Neef, A. and Grayman, J.H. (eds) *The Tourism-Disaster-Conflict Nexus.* Emerald Publishing: Bingley, pp. 87–104.

Weeden, C. (2015) 'Legitimization through corporate philanthropy: A cruise case study', *Tourism in Marine Environments,* 10, (3–4): 201–210.

Media sources/websites

Adams, D. (2014) 'Neglected islanders resist plan for Haiti tourism revival', 6 April. Available at: www.reuters.com/article/us-haiti-tourism/neglected-islanders-resist-pla n-for-haiti-tourism-revival-idUSBREA3506V20140406 (accessed 1 January 2019).

Ambrose, D. and Majeed, K. (2018) 'Typhoons and tycoons: Disaster capitalism in the Philippines', *Al Jazeera,* 16 August. Available at: www.aljazeera.com/indepth/features/ typhoons-tycoons-disaster-capitalism-philippines-180816065729201.html (accessed 1 January 2019).

Ayala Corporation (2018) *Annual Report 2017.* Available at: www.ayala.com.ph/sites/defa ult/files/disclosures/AYALA%20CORPORATION%20SEC%2017A%20ENDING% 20DEC.%2031%2C%202017.pdf (accessed 13 January 2019).

Bangkok Post (2019) 'The Moken fight for more space', 18 February. Available at: www.bangkokpost.com/thailand/special-reports/1630670/the-moken-fight-for-mor e-space (accessed 10 September 2020).

Charles, J. (2020) 'Bill Clinton once enjoyed a bright legacy in Haiti. Then the 2010 earthquake struck', *Miami Herald,* 10 January. Available at: https://pulitzercenter. org/reporting/bill-clinton-once-enjoyed-bright-legacy-haiti-then-2010-earthquake-struck (accessed 6 June 2020).

Chery, D. (2014) 'Human rights organizations: widespread abuse and police brutality in Haiti's Île à Vâche', *News Junkie Post,* 8 April. Available at: http://newsjunkiepost. com/2014/04/08/human-rights-organizations-widespread-abuse-and-police-brutalit y-in-haitis-ile-a-vache/ (accessed 6 June 2020).

Conserva, L.H. (2014) 'Ayala-Sarrosa joint venture commits P114M for Sicogon farmers', 19 November. Available at: www.bworldonline.com/content.php?section=Econom y&title=ayala-sarrosa-joint-venture-commits-p114m-for-sicogon-farmers&id=98132 (accessed 3 January 2019).

Daut, M. (2020) 'When France extorted Haiti: the greatest heist in history', *The Conversation,* 30 June. Available at: https://theconversation.com/when-france-extor ted-haiti-the-greatest-heist-in-history-137949 (accessed 6 August 2020).

Golden, F. (2016) 'Cruise ship turns away from Haiti amid protests', *USA Today,* 20 January. Available at: www.usatoday.com/story/travel/cruises/cruiselog/2016/01/20/ royal-caribbean-ship-drops-haiti-amid-protests/79056806/ (accessed 6 June 2020).

Ives, K. (2014) 'Île à Vâche under siege', *Haiti Liberté* 7 (36). Available at: https://cana da-haiti.ca/content/%C3%AEle-%C3%A0-v%C3%A2che-under-siege (accessed 6 June 2020).

Jubis, S. (2015) 'Garifuna communities of Honduras resist corporate land grabs', Council on Hemispheric Affairs (COHA), 23 September. Available at: www.coha. org/garifuna-communities-of-honduras-resist-corporate-land-grabs/ (accessed 14 January 2019).

Kushner, J. (2015) 'Paradise is overbooked', *Foreign Policy,* 15 April. Available at: http s://foreignpolicy.com/2015/04/15/paradise-overbooked-haiti-land-earthquake-tour ism/ (accessed 8 June 2020).

Laiko (2018) 'Sicogon farmers protest against land grabbing and disaster capitalism', Council of the Laity of the Philippines (Laiko), 7 December. Available at: www. cbcplaiko.org/2018/12/07/sicogon-farmers-protest-against-land-grabbing-and-disast er-capitalism/ (accessed 2 January 2019).

Lakhani, N. (2020) 'Fears growing for five indigenous Garifuna men abducted in Honduras', *The Guardian*, 23 July. Available at: www.theguardian.com/global-development/202 0/jul/23/garifuna-honduras-abducted-men-land-rights (accessed 21 August 2020).

Marshall, N. (2015) 'Your holiday can help: Vanuatu and Nepal appeal for tourists to return', *The Guardian*, 13 August. Available at: www.theguardian.com/travel/2015/aug/14/ your-holiday-can-help-vanuatu-and-nepal-appeal-for-tourists-to-return (accessed 23 May 2018).

Panchang, D., Bell, B., and Field, T. (2012) '"The Super Bowl of disasters": Profiting from crisis in post-earthquake Haiti', 19 February. Available at: https://hcvanalysis. wordpress.com/2012/02/19/the-super-bowl-of-disasters-profiting-from-crisis-in-post -earthquake-haiti/ (accessed 19 August 2020).

Philippine Daily Inquirer (2019) 'Ayala Land, resorts developers ordered to stop dev'ts in disputed land on Sicogon', 22 March. Available at: https://newsinfo.inquirer.net/ 1098810/ayala-land-resorts-developers-ordered-to-stop-devts-in-disputed-land-on-si cogon (accessed 3 June 2020).

Schnabel, C. (2017) 'UN honors Zobel de Ayala as pioneer of sustainable develop- ment', 14 September. Available at: www.rappler.com/business/182183-united-na tions-jaime-augusto-zobel-de-ayala-pioneer-sustainable-development-goals (accessed 3 January 2019).

Schuller, M. and Tsu, J. (2014) 'Haiti "open for business," like it or not', 17 June. Available at: www.wilderutopia.com/international/earth/land-grab-in-paradise-haiti s-ile-a-vache-fights-back/ (accessed 6 June 2020).

Smillie, S. (2019) '"It's worse than the tsunami": The sea nomad village devastated by fire', *The Guardian*, 26 February. Available at: www.theguardian.com/global-development /2019/feb/26/moken-sea-nomad-village-devastated-by-fire (accessed 10 September 2020).

Walker, J. (2010) 'Labadee – Royal Caribbean's deal with the devil', *Cruise Law News*, 28 January. Available at: www.cruiselawnews.com/2010/01/articles/worst-cruise-li ne-in-the-world/labadee-royal-caribbeans-deal-with-the-devil/ (accessed 6 June 2020).

6 Tourism, dispossession and erasure in conflict zones and post-conflict contexts

The notion that tourism, militarism and conflict can go hand in hand may seem rather remote to most tourists. Yet, recently, a number of studies have looked into tourism's controversial entanglements with war, conflict and militarism (Weaver, 2011; Gonzalez, 2013; Lisle, 2016). War zones have often produced a particularly problematic brand of tourism, especially in tourist enclaves prepared exclusively for soldiers during their 'rest & recreation' phases, such as Thailand's Pattaya beach area for the US military during the Vietnam war (cf. Lisle, 2016). Today, the Thai military holds several properties across the kingdom that are used as recreation facilities for its armed forces and their relatives. In her book *Securing Paradise*, Gonzalez (2013) shows how militarism and tourism have been interwoven in Hawai'i and the Philippines, thereby empowering the United States to advance its geopolitical and economic interests in the Pacific. She argues that "tourism – with its structuring ideas and practices of mobility and consumption – is the perfect partner to militarism's claims to security" (Gonzalez, 2013, p. 4).

While tourism has a potential role to play in rehabilitating areas devastated by genocide, war, and other types of armed conflict (e.g., Alluri, 2009; Causevic and Lynch, 2011; Novelli, Morgan and Nibigira, 2012; Neef and Grayman, 2018), in many cases the military – aiming to sustain its economic interests, maintain its territorial control and keep its soldiers busy in more peaceful times – has played a much more controversial part, including the forceful taking of land from former adversaries in civil conflicts. The contentious role of the military in conflict and post-conflict tourism development will be more closely examined hereafter for the cases of Myanmar (formerly Burma), Sri Lanka and Bangladesh. The final case examines how tourism is instrumentalised in the on-going Israeli-Palestinian conflict.

Tourism, displacement and the military in Myanmar (Burma)

For more than half a century (from 1962–2011), Myanmar (formerly Burma) was subjected to authoritarian rule. Following an uprising in 1988 that was brutally crushed by the military, the country was ruled for ten years by the so-called State Law and Order Restoration Council (SLORC), until it was replaced by the State Peace and Development Council (SPDC) in 1997 (Henderson, 2003). One of the first major policy actions of the SLORC post-1988 was to relocate hundreds of thousands of people from the then-capital Rangoon, later renamed Yangon by the regime (Rhoads, 2018). While the primary objectives of these forced relocations were to establish 'law and order', build new infrastructure, and erase "sites and communal memories associated with past resistance against the state" (Seekins, 2011, p. 161), they also had the welcome side effect of beautifying and securing the city to appeal to the foreign tourist market (Parnwell, 1998).

Rapid tourism development was high on the agenda of the post-1988 military regime for both economic and political reasons; it was in urgent need for foreign currency and international recognition (Henderson, 2003; Clifton, Hampton and Jeyacheya, 2018). As the SLORC lacked the financial resources to build a viable tourism industry, it sought foreign investment and resorted to forced labour (Parnwell, 1998). Hundreds of thousands of people (including children, pregnant women and the elderly) were coerced by the military junta to build airports, upgrade railways and restore roads as well as temples, palaces and other potential tourism sites, especially in preparation for the Visit Myanmar Year (1996–1997) campaign (Yokota, 1996; Parnwell, 1998). Shackled slave labourers were witnessed working on an airport extension in today's Pathein, the capital city of the Ayeyarwady Division (Mahr and Sutcliffe, 1996). Henderson (2003) reported that the number of airports in Myanmar increased from 43 in 1988 to 66 in 2001. The construction of the Ye-Tavoy railway – enabling tourist-access to southern Myanmar – involved forced labour of over 200,000 people and cost the lives of about 300 workers due to the appalling working conditions (Mahr and Sutcliffe, 1996).

Confiscation of land and forced resettlements were other important components of the tourism strategy of the military junta and its cronies (see Table 6.1). In coastal areas of the then Irrawaddy Division and Arakan State, villagers lost their farmland, gardens, coconut groves, homes and guesthouses to large-scale beach hotel development between 1990 and the early 2000s (Thitsar, 2013; Soe and Ko, 2014). Other areas that were identified by the regime as major tourist centres in the 1990s were Mandalay, Bagan and Taung-gyi near Inle Lake, where evictions and state-led land grabbing were also common. Members of the military elites and their cronies have had major stakes in those tourism development projects, and for many years foreign tourists were only allowed to stay in government-owned or government-licensed accommodation, while homestays in communities were prohibited (Henderson, 2003; Zhou, 2005).

In Bagan, one of the most stunning cultural landscapes in Myanmar, up to 5,000 people – many of whom had been traditional caretakers of cultural heritage sites – were forcibly removed from the grounds of hundreds of ancient temples and pagodas in the early 1990s (Pilger, 1996). Disregarding advice from international experts and even its own tourism agency, the SLORC believed that the relocation would help them secure a nomination for

Table 6.1 Prominent examples of tourism-related land confiscations and displacements in Myanmar

Township (division, region or state)	Year/period	Affected area/ population	Impact on communities
Chaung Tha (Irrawaddy Division; Ayeryawady Region)	1989–1997	615 acres (249 hectares)	Confiscation of farmland and gardens with inadequate compensation
Bagan (Mandalay Division; Mandalay Region)	1990	4,000–5000 people	Forced relocation to infertile land, confiscation of farmland, bulldozing of houses
Ngwe Saung (Irrawaddy Division; Ayeryawady Region)	Early 1990s–2000	Seven villages	Confiscation of farmland with inadequate compensation, relocation of communities
Thandaung (Karen State; Kayin State)	1995	Up to 200 ethnic minority villages	Ethnic minority groups forcibly relocated to live in model tourist villages in the periphery of Yangon
Lampi Marine National Park (Tanintharyi Region)	1996	400 indigenous seafaring people, 205km^2	Indigenous Moken deprived of their livelihoods as subsistence fishers through the declaration of on-fishing zones, relocation of at least one community
Ngapali (Arakan State; Rakhine State)	2000	36 acres (15 hectares)	Confiscation of farmland without compensation, dismantling of village guesthouses
Nyaungshwe (Inle Lake, Shan State)	2012–2013	600 acres (243 hectares)	Confiscation of farmland with inadequate compensation, seven villagers charged with obstruction
Tada-U (Mandalay Division; Mandalay Region)	2012–2014	2,000 acres (809 hectares)	Forced land acquisition at inadequate land prices

Source: Pilger, 1996; Henderson, 2003; Thitsar, 2013; Soe and Ko, 2014; Aye, 2015; Clifton, Hampton and Jeyacheya, 2018.

a UNESCO listing in 1996 but their proposal was turned down at the time (Kraak, 2017; Tha, 2019). Subsequently, the regime opened up the landscape for unfettered tourism development. The so-called 'Property Zone' of the Bagan cultural landscape is now dotted with a total of 85 hotels built by military cronies and international hotel chains during the regime of the military junta until 2011 and the military-backed transitional administration of U Thein Sein from 2011 to 2016. Just before handing over the country's government to the elected National League for Democracy (NLD) under Aung Sang Suu Kyi in March 2016, U Thein Sein's outgoing administration approved 42 new hotel construction projects in the 'Property Zone' (Tha, 2019). Some of the hotels – such as the Hilton Bagan and the Aureum Palace Hotel – even have century-old temples within their compounds (Tha, 2019). Yet, as the new government was finally successful in its bid for Bagan's UNESCO world heritage status in 2019, it had to commit to moving all hotels from the 'Property Zone' to a specially designated 'Hotel Zone' by 2028 (Chau and Htet, 2019).

Land acquisitions by the state for tourism as a 'public purpose' are facilitated by a hybrid legal framework (see Box 6.1) that combines elements from the colonial era, such as the 1894 Land Acquisition Act, with investment-friendly legislation and repressive new land laws enacted under the military regime's successor government (2011–2016). Particularly the Farmland Law and the Vacant, Fallow, and Virgin Law (VFV) – both promulgated in 2012 – provide the regulatory basis for further land grabs and make existing ones lawful (Transnational Institute, 2013; Scurrah, Hirsch and Woods, 2015). The Farmland Law only recognises property rights on officially registered land with valid land use certificates (LUCs) which are unattainable for most smallholder farmers, as the registration process is either extremely slow or non-existent (Oberndorf, 2012). Even if a farmer has access to land registration procedures, the burden of proof concerning the pre-existence of the right to use the land is on her or him and s/he also has to demonstrate 'proper' use. The VFV regards all untitled lands – including forestlands and land that is actually being used – as 'wasted assets' and gives the government the right to redefine them as 'vacant, fallow or virgin' and subsequently reallocate those lands to foreign or domestic investors, including for tourism purposes (Carter, 2015). Land confiscation may affect swidden land, communally managed pastures, village forests and fishponds, which are crucial for rural people's food security and livelihoods but are neither formally registered nor mapped.

Box 6.1 National land legislation enabling land confiscation by the state and private investors in Myanmar

The Land Acquisition Act (1894)

Hailing from British colonial rule, the act continues to be the major legal instrument used by the Myanmar government to confiscate land; has provisions for appropriate processes of land acquisition including compensation procedures, but in reality, these were mostly disregarded under military rule.

2008 Constitution

Identifies the state as the ultimate owner of the land in Myanmar; gives the government the right to acquire land from its citizens against their will.

2012 Farmland Law

Stipulates that use rights on farmland need to be registered, which gives the owner a land use certificate (LUC) representing tenured title; creates a situation where anyone without an LUC no longer possesses legal rights to use land and can be evicted; creates the Farmland Administration Body (FAB), chaired by the Minister of Agriculture and Irrigation, which has sole power to allocate land use rights; does not subject the FAB to the reach of the judicial system, which means that decisions made by FAB are final and cannot be appealed in a court.

2012 Vacant, Fallow, and Virgin Lands Management Law (VFV Law)

Reclassifies 'unoccupied' land as 'vacant, fallow and abandoned' and makes it available as farmland; declares 'reserved forests, grazing ground and fishery pond land' and 'uncultivated' land as 'virgin land' and makes it available for a variety of alternative uses; allows the Central Committee to allocate such 'vacant, fallow or virgin' land for use in large-scale agriculture, mining, tourism and other purposes permitted by the government; permits the allocation of up to 50,000 acres (about 20,234 hectares) to foreign investors in the form of long-term leases.

2012 Foreign Investment Law

Allowed foreign investors to lease private land with an initial investment term of 30 years, twice extendable for periods of 15 years; offered tax breaks to foreign investors; enabled them to establish businesses without the need for local partners.

2015 Myanmar Investment Law

Combines the 2012 Foreign Investment Law and the 2013 Myanmar Citizens Investment Law; alters the mandate of the Myanmar Investment Commission (MIC); adds some nominal human rights protections to future foreign investment projects.

Source: Oberndorf, 2012; Carter, 2015; Scurrah, Hirsch and Woods, 2015; Neef, 2016

One of the particularities of Myanmar is the presence of 'Special Regions' or autonomous zones (Than, 2016). In the first decade (1989–1999) of the military regime, ceasefire agreements were signed with 26 armed groups, with 13 of them being subsequently granted Special Region status (Than, 2015). Most of these Special Regions, including Kachin State, Karen (or

Kayin) State and Shan State, are home to diverse Indigenous groups, referred to as 'ethnic nationalities' by the Myanmar government. Tourism development in these sensitive borderland areas of northern and northeastern Myanmar has been employed by the military regime to tighten its control over territory and people. In Shan State, for instance, the then leader of the military junta, General Than Shwe, ordered the destruction of Kengtung Palace in late 1991 (Hurng, 2020). The palace had been built by a Shan prince in 1906. At the time when the British Empire invaded the area, Kengtung was the largest and most significant Shan principality in present-day Shan State. After blowing up the palace, the military regime leased the land to a private company for a period of 70 years and a sum of US$138,000 to build the Amazing Kyaing Tong Resort (Hurng, 2020). Emboldened by the nascent democratisation process in Myanmar, local community leaders collected support for a petition to the State Counsellor Aung Sang Suu Kyi in 2019 demanding the return of the land to the Shan people. Yet the Vice-Minister of Hotels and Tourism was quoted saying that it was impossible to transfer the land back to the community before the end of the lease, i.e. in 2062 (Hurng, 2020).

Cross-border casino tourism has become a particularly prominent feature in Myanmar's Special Regions (e.g. Than, 2015). Nearly all customers come from neighbouring China and Thailand where gambling is illegal. While gambling was prohibited even for foreigners in Myanmar until 2019, Special Regions can set up their own rules and have legalised gambling for cross-border tourists. As exemplified in Box 6.2, casino development often goes along with large-scale land acquisitions and forced relocations. Similar processes have been recorded in Laos and Cambodia where international casino tourism has also experienced a boom (Than, 2015; Sims, 2017).

Box 6.2 International casino tourism in the Myanmar-Thai borderlands

Casino tourism has become pervasive in Myanmar since the early 2000s, often combined with the development of megaresorts and even entire new city projects. One such megaresort – embedded in an international 'IT industrial city' project – has been planned since 2017 along the Myanmar-Thai border by the Yatai International Holding Group which is registered in Hong Kong and headquartered in Bangkok. The multinational corporation claims to have made an initial investment of US$500 million and anticipates an overall investment of a staggering US$15 billion for villas, an international airport, casinos and various other tourist attractions, including an 'ethnic shopping mall' and a 'Safari World', on an area of 12,000 hectares (29,653 acres), branded as the 'Myanmar Yatai Shwe Kokko Special Economic Zone'. In September 2017, the company signed a partnership agreement with Colonel Chit Thu, the commander of the Karen Border Guard Force (KBGF), an armed group backed by the Myanmar military.

The Shwe Kokko area had been at the centre of a long-standing conflict between the Burmese government army and the Karen National Union (KNU) who controlled most of this borderland area until the mid-1970s. The last stronghold of the KNU was seized by the Myanmar military in 1995 with support from the newly formed Democratic Karen Buddhist Army (DKBA) which had split away from the KNU. In 2010, the DKBA was transformed into the KBGF under the Myanmar Army. For the Yatai International Holding Group, signing an agreement with the KBGF commander meant to ensure the protection of its large-scale development project against community protests and obstructions.

Local communities in Shwe Kokko learned about the new project in early 2017, when the KBGF authorities informed them in a meeting in the village temple that their lands would be confiscated for a 'new city expansion' development, without providing any details of the nature of the project. Community members had no choice but to accept the promised compensation of about US$400 per hectare of confiscated land and between US$3,000 and US$5,000 for their houses, depending on their size and quality. While local people were told that the project would provide employment opportunities that would make up for their loss of livelihoods derived from farming, thousands of Chinese workers have been brought in for the construction of tourism and other infrastructure.

Reportedly, the investors have obtained a land use permit for 70 years, with a possible extension to 99 years, and are exempted from income tax for the first ten years of operation. Both the duration of the lease and the duration of the tax break are higher than officially permitted under Myanmar law. There is also confusion over the legally approved scale of the project. According to local media, the Myanmar Investment Commission (MIC) approved only US$22.5 million in investments for the construction of nearly 60 villas on slightly over ten hectares of land. Claims by the company that it is part of the Chinese government's Belt and Road Initiative and a trilateral China-Thailand-Myanmar economic corridor have been called into question, as Chinese authorities have recently cracked down on Chinese-operated casino businesses in Laos and Myanmar.

While Myanmar promulgated a new Gambling Law in 2019, legalising casino operations that cater for foreigners only, the Myanmar government reportedly denied that it permitted gambling activities in the Shew Kokko area. Yet it seems obvious that the casino business will have a twofold advantage for the Myanmar military, by solidifying their control of this strategic border area and by reaping their share of the expected profits from non-transparent casino operations.

Source: Nyein, 2019; KPSN, 2020; Lwin, 2020

Casino cities have been at the centre of conflict between the central government and separatist groups. In 2009, the Myanmar military attacked the Myanmar National Democratic Alliance Army (MNDAA) in Laukkaing city, the capital of the Kokang self-administered zone at the border with China, which hosts several casinos (International Crisis Group, 2019). The attack forced the long-standing commander of the city to flee across the border into China with his soldiers and an estimated 37,000 displaced civilians (Kramer, 2009). Following the expulsion of the MNDAA, the military junta installed a pro-government Border Guard Force to administer the casino city (International Crisis Group, 2019). It is obvious that for Myanmar's ruling elite, casino tourism has played a dual role of (1) extending their influence over ethnic nationalities in semi-autonomous restive states and (2) cashing in on the enormous profits that can be made from the gambling business.

Military-controlled tourism industry in post–conflict Sri Lanka

Celebrating 70 years of independence from the British Empire, Sri Lanka was named Lonely Planet's top destination for 2019. According to the travel advisor's website its "endless beaches, timeless ruins, welcoming people, oodles of elephants, rolling surf, cheap prices, fun trains, famous tea and flavourful food make Sri Lanka irresistible" (Lonely Planet, 2020, n.p.). For obvious reasons, there was no mention of the 2004 Indian Ocean Tsunami that brought devastation to the northern and eastern parts of the island country nor of the horrors of the 26-year civil war between the Sinhalese majority and the Liberation Tigers of Tamil Eelam (LTTE) that came to an end in 2009. Since the defeat of the LTTE, its former strongholds in the northern and eastern provinces have remained heavily militarised. Around three-fourths of Sri Lanka's 200,000 armed forces remain stationed in the northern province, where they build Buddhist temples and schools and operate the largest hotels and golf courses in the Jaffna area, often on land confiscated during or after the civil war (Draper, 2016).

According to the UK-based Sri Lanka Campaign for Peace and Justice (SLCPJ, 2019), many tourism companies in Sri Lanka are linked with individuals and organisations implicated in war crimes or serious human rights violations, while others are deemed problematic because they have been involved in the white-washing of human rights abuses. The northeastern sites of the final stages of the civil war have been transformed into places of triumphalism through the construction of military monuments as symbols of domination, often associated with the eviction and repression of the Tamil minority population which has been deemed a defeated people by majority Sinhalese since the Sri Lankan army's victory over the LTTE (Seoighe, 2016).

The Sri Lankan military owns a wide range of tourist facilities and launched its own resort brand, Laya (a Sanskrit word meaning 'rest and repose'), in 2012 (Seoighe, 2016). The Sri Lankan Navy offers whale watching tours and runs ferry services, while the Sri Lankan Air Force owns a commercial airline – Helitours – which according to its own website – "is the prime domestic airline operating in Sri Lanka connecting all the air fields within the country" and "stands as a dominant player in promoting the tourism in the country". Many resorts, hotels, restaurants and memorial sites were built on land that was forcefully taken from its previous owners. Hence, the 'corporate military' in Sri Lanka uses the booming tourism sector in the country to provide employment opportunities for its oversized forces and thereby controls a major share of the country's economy, while dispossessing many Tamil citizens, cutting off their livelihood resources and systematically trying to erase their culture and collective memory, a process that is understood by many Tamils as a continuous structural genocide (Seoighe, 2016).

In recent years, some areas around Arugam Bay on Sri Lanka's southeastern coast have piqued the interest of the economic arm of Sri Lanka's military. The area came into the international spotlight following the devastation of the 2004 Indian Ocean Tsunami, when a swiftly drafted resource development plan with support from USAID, World Bank, and the Asian Development Bank was aimed at transforming the coastal township into a high-end boutique tourism destination, much to the delight of hotel owners and investors who had previously been in conflict with local fishing families over the use of coastal resources (Klein, 2007). A notorious and widely reported case of tourism-related land grabbing is 'Lagoon Cabanas', a luxury resort in Panama, a small town with a surfing beach just south of Arugam Bay. It is part of the 'Malima' chain, owned by the Sri Lankan Navy. 'Lagoon Cabanas' is part of a wider tourism development programme in the area which, according to a report by the UK-based NGO Oxfam and records of the UN Refugee Agency (UNHCR) and Human Rights Watch, has seen the forced relocation of over 350 families from their homes without proper compensation (Piyadasa, 2016; Ratnayake and Hapugoda, 2017). In July 2010, one year after the official end of the civil war – masked and heavily armed men wielding weapons began attacking the villages and expelled the local residents. Following this incident, the land was confiscated by the army. Several residents who complained about the confiscation to a People's Tribunal in Colombo were subsequently threatened by the police. Many have continued their fight for justice while living in displacement camps and temporary shelters (International Tribunal on Evictions, 2017).

A guide to ethical tourism in Sri Lanka, published by the Sri Lanka Campaign for Peace & Justice (SLCPJ) in October 2018, lists 21 hotels and resorts, four restaurants, three golf courses, two whale watching tours, two airlines, one ferry service, one diving centre, and one nature reserve that are all owned by the Sri Lankan military (Sri Lanka Campaign for Peace & Justice, 2018). A continuously updated website provides potential travellers with information about which tourism companies and facilities they should avoid if they want to make sure that their holiday does not benefit human rights abusers (see Box 6.3 for selected examples). Cases have also been examined and reported by Sri Lanka's National Fisheries Solidarity Movement (NAFSO), the Sri Lanka Nature Group as well as the Society for Threatened Peoples (STP, 2015).

Box 6.3 Ethical tourism in Sri Lanka – Examples of places and services to avoid

The following are examples of touristic places and services that are deemed unethical to visit, according to the Sri Lanka Campaign for Peace and Justice.

'Thalsevana Holiday Resort' is a military-run holiday resort functioning "under the Security Forces Headquarters" on the northernmost beach of Kankesanthurai, Jaffna. The resort is located directly inside the intensely militarised High Security Zone (HSZ), a vast area of land that has been appropriated by the Sri Lankan government from private owners in successive waves from the 1990s onwards. Following the end of the war in 2009, at least 2,830 hectares were newly acquired. The HSZ was initially built to "blockade against Tiger resupply ships," but has continued to operate on a vast scale despite the defeat of the LTTE. The expelled Tamil inhabitants – including many fishermen – continue to protest in hope of their land being returned by the government. In April 2017, a small portion of several hundred acres was returned to its owners, many of whom found their homes destroyed or in a state of disrepair. At the handover ceremony, the Sri Lankan army commander issued a chilling warning to those present, stating that "[just] like we grant you these houses and lands, we are able to take it back again."

'Malima Hospitality Services (MHS)' is a chain of resorts and attractions owned by the Sri Lankan Navy. Its portfolio includes hotels which have been built on the lands of private citizens who have been forcibly evicted, including in Panama village, south of Arugam Bay.

'Nature Park Holiday Resort' is a venture run by the Sri Lankan Navy. It is located inside 'Chundikkulam Bird Sanctuary' on the southern-eastern tip of the Jaffna peninsula, an area of land that was seized by the military at the end of the war in 2009. Reports suggest that many of the Tamil landowners who had originally occupied the land have been refused access to it.

Wildlife sanctuaries in the Mullaitivu district were established following the end of the conflict, as a means of attracting tourists to these areas. 'Kokkilai Lagoon Sanctuary', which covers an area of forest formerly occupied by the LTTE and which was the scene of heavy fighting during the end of the war, is one such venture.

'Eagles' Golf Links' and 'Eagles' Heritage Golf Course' are golf courses owned and operated by the Sri Lankan Air Force in Trincomalee and Anuradhapura. The Trincomalee course was opened in 2012 in a ceremony led by former Defence Secretary Gotabaya Rajapaksa, an alleged war criminal. According to its website, "proceeds [from the golf course] are utilized for the benevolence of [Air Force] personnel and their family members."

Source: www.srilankacampaign.org/ethical-tourism/avoid/
(accessed 2 January 2019)

All studies are unanimous in their conclusion that the ongoing securitisation, militarisation and 'Sinhalisation' in the disguise of a booming tourism industry in Sri Lanka's northern and eastern provinces risks undermining the prospects of lasting peace in the country (cf. Seoighe, 2016; Ratnayake and Hapugoda, 2017). Unless confiscated lands are returned to their rightful owners and local residents can benefit more substantively from the tourism sector and other economic activities, community resentment against the various branches of government is not likely to abate. Yet hopes for such a process of land restitution, transitional justice and national reconciliation have been dashed by the landslide victory in the August 2020 parliamentary elections of the Sri Lanka Podujana Peramuna (SLPP) party which focused its campaign on a Sinhala Buddhist nationalist ideology.

Military securitisation of domestic tourism in the Chittagong Hill Tracts of Bangladesh

The Chittagong Hill Tracts (CHT) – which obtained their name under British colonial rule in 1860 – are a region in the southeastern corner of Bangladesh, inhabited by the indigenous Jumma people, which comprise several subgroups, such as the Chakma, Mro and Marma. The Jumma have a long history of resistance against external forces, dating back to colonial times. In 1997, a twenty-year long armed conflict between the Government of Bangladesh and the United People's Party of the Chittagong Hill Tracts and its armed wing, the Shanti Bahini, over the issue of autonomy and land rights of the Jumma people finally came to an end. During this long civil conflict, the Bangladesh military had burned homes of the Jumma, carried out mass killings and strategically placed hundreds of thousands of Bengali settlers on Jumma land near their military bases (IWGIA, 2012; Ahmed, 2017). This strategy changed the demography of the CHT; where the Jumma people made up 98 per cent of the population in 1947, their share had dropped to 51 per cent in 1991 (Mohsin, 1997, cited in Ahmed, 2017).

Twenty years after the 1997 Peace Accord, the Government of Bangladesh announced 2016 as the 'Tourism Year' and started a three-year promotional campaign for tourism, including in the CHT which has become a popular destination for domestic tourists (Ahmed, 2017). Under the Bangladesh Army Welfare Trust the military maintains a fair share of the tourism industry, owning Radisson Hotels in the capital Dhaka and the port city of Chittagong along with deluxe golf courses. It also holds a range of tourism facilities in the CHT. It is therefore not surprising that the military has not reduced its presence in CHT despite the Peace Accord, which promised to demilitarise the region (Ahmed, 2017). According to the IWGIA (2012), one third of the country's armed forces remained in CHT, despite its low population density. Thereby, the military plays a dual function of running a major portion of the tourism sector in the CHT as well as securitising the region for visitors from other parts of the country. When middle-class Bengalis visit the area as tourists, their major concern is their own safety, hence they gladly accept a high level of securitisation by the military and the associated narrative of preserving Bangladesh's sovereignty and border security (Ahmed, 2017).

According to Jumma advocacy organisations, at least 688ha of land has been allocated to build tourist resorts, resulting in the eviction of more than 700 Jumma families from 26 villages (Chakma and Chakma, 2015, cited in Ahmed, 2017). The process of forced land acquisition involves powerful developers with links to state authorities asking the District Commissioner to declare the land as 'khas' which means 'state-owned' (Adnan and Dastidar, 2010, p. 141). The 'khas' status renders occupancy by Indigenous people illegal and frees the land for acquisition under private property laws (Chakma, 2017). Cultural erasure is also present, as local names for important places are changed into Bengali terms (Ahmed, 2017). One example of tourism development that has involved dispossession, eviction and other human rights violations is the Nilgiri Resort in the Bandarban Hill District, a luxury tourist resort on a mountain hilltop established and run by the Bangladesh military (IWGIA, 2012). This development entailed the forceful displacement of about 200 Mro and Marma families by the military, which destroyed villagers' orchards that were their main livelihood source as well as several shops and a school (Chakma, 2017; Ahmed, 2017). Another military-run tourism facility is the nearby Nilachar Lodge, four kilometres from Bandarban town (IWGIA, 2012).

A Jumma activist and blogger lists a number of other tourism businesses run by the Bangladesh military in CHT. These are:

- Neel Giri Tourist spot in Bandarban district (acquisition of approximately 243ha (600 acres) by the Bangladesh Army);
- Dim Pahar in Bandarban district (acquisition of approximately 202ha (500 acres) by the Bangladesh Army);
- Lake Paradise at Kaptai, Rangamati district (run by the Bangladesh Navy);
- Jibtoly Resort at Kaptai, Rangamati district (run by the Bangladesh Army);
- Agitator at Baghaichari, Rangamati district (run by Border Guards Bangladesh);
- Heritage Park at Changi Bridge, Khagrachari district (run by Ansar – a paramilitary auxiliary force responsible for the preservation of internal security and law enforcement - and Village Defence Party – an enforcement unit at the level of individual villages and urban towns); and
- Rui Lui valley on Sajek Hill, Rangamati district (run by the Bangladesh Army).

(Tripura, 2016)

Ironically, an important source of capital for the business wing of the Bangladesh military, which runs the extensive tourism facilities, are funds earned from its UN peacekeeping missions (IWGIA, 2012). The World Bank is also playing its part through a proposed US$360 million sub-regional connectivity project between India, Nepal, Bhutan and Bangladesh (World Bank, 2016). The Ministry for CHT Affairs (MOCHTA) which was established through a provision under the CHT Accord with a mandate of implementing the Accord has become more involved in promoting tourism and cultural aspects related to the CHT and engaging in joint ventures with the Ministry of Civil Aviation and Tourism (Ahmed, 2017). Future plans are to attract more foreign tourists to the CHT, who thus far have been deterred by the various travel warnings that are in place due to the tense security situation.

The majority of Jumma people do not benefit from this military-led tourism development, as expressed by the following statement of a Jumma student activist:

> You don't need a gun to kill me … You can kill me just through doing your form of 'development' … The kind of development you are doing through tourism has evicted Jumma people from their land … We want an end to this process. For us, each of the tourist spot [*sic*] is a weapon against us.
>
> (cited in Ahmed, 2017, p. 121)

To resolve the numerous cases of land disputes in the Chittagong Hill Tracts, the CHT Land Dispute Resolution Commission Act was promulgated in 2001, with further amendments by the government in October 2016. By the end of 2017, the reconstituted land commission had received more than 22,800 complaints, but has yet to solve any of the land conflicts due to a lack of human and material resources and the absence of any supplementary rules to the Act that would guide its implementation (Chakma, 2018). Meanwhile, in 2017, a total of 141 Indigenous human rights defenders and indigenous villagers were reportedly arrested or detained by government forces (Kapaeeng Foundation, 2018).

In some cases, resistance by local people against destructive tourism developments has been successful, particularly when organised with the help of civil society groups in CHT and the capital Dhaka. Ahmed (2017) reports the case of the Alutila Special Tourism Zone which the Bangladesh Economic Zone Authority (BEZA) under the Prime Minister's Office had planned on about 283 hectares of land in Khagrachari district. Although the government marked it as '*khas*' (state) land, the area has been inhabited by more than 500 indigenous families in 21 villages for generations, which would have faced eviction (*New Age*, 2016). The families joined forces with two Jumma political groups and activist networks in Dhaka to protest the government's plan, which was eventually cancelled by MOCHTA in October 2016 (*The Daily Star*, 2016). Nevertheless, the indigenous people of the CHT continue to live in constant fear of eviction from their homes and ancestral lands.

Even the COVID-19 pandemic did not prevent tourism developers from further grabbing land from Indigenous communities. In October 2020, villagers from eight Mro communities in Bandarban Hill District submitted a memorandum to the country's Prime Minister, alleging that officials of a welfare organisation and a large business entity have jointly encroached on their farmland, village forest, cremation grounds and fruit orchards during the pandemic to construct a five-star hotel that would lead to the eviction of hundreds of families from their ancestral territory (Barua, 2020).

Touristification, displacement and erasure in Jerusalem amidst the Israeli–Palestinian Conflict

Following the six-day Arab-Israeli war in June 1967 Israel annexed East Jerusalem, which had been under Jordanian rule, together with an additional 60km² of land in surrounding areas of the West Bank (henceforth 'East Jerusalem'). The annexation significantly expanded the municipal borders of the city. Most of the confiscated land had previously been under private Palestinian ownership but Israel froze the land regularisation and registration processes in East Jerusalem and the West Bank which had been established under Jordanian rule (Ir-Amim, 2015). This freeze related solely to non-Jewish property owners; in the new neighbourhoods built on confiscated land, the registration of land in the name of the new (Jewish) owners was implemented in full. As of 2015, approximately one-half of the land in East Jerusalem was not registered in any form (Ir-Amim, 2015). Palestinian residents face various challenges regarding their rights to residence and construction of homes, including the need to prove landownership, complex processes of obtaining building permits and the expansion of Jewish neighbourhoods (Ir-Amim, 2012). This legal situation provides the backdrop for a long-standing conflict involving tourism, heritage, settlement and land rights in the village of Silwan.

The village of Silwan has a population of about 40,000 Palestinians and is located only a few hundred metres from the walls of Jerusalem's Old City and the Temple Mount (known among Muslims as Haram al-Sharif). Residents of the Al-Bustan neighbourhood have been involved in a decades-long struggle that began in the 1970s with the Israeli government's plan of building a national park in the area (Laskin, 2013; Ma'an, 2017). Since 1997 – when the area was designated as an 'open public area' (despite a proliferation of private housing) – no building permits have been issued in Al-Bustan (Ir-Amim, 2012).

The municipality began issuing demolition orders and indictments to Palestinian households in Al-Bustan in 2005 as part of the Israeli authorities' plan to establish an archaeological touristic park called 'The King's Valley' in Silwan and around the 'Holy Basin' which includes several Christian and Muslim holy sites (Bronner, 2008; Ir-Amim, 2012; Ma'an, 2017). The first two homes in Al-Bustan were destroyed later that year (Gadzo, 2017). Yet the planned evictions of some 100 Palestinian families met international criticism which contributed to temporarily stalling the process (Ma'an, 2017). An alternative plan for the neighbourhood developed by the residents of Al-Bustan with the help of local architects and planners was rejected by the District Committee in 2009 (Ir-Amim, 2012).

In the same year, the Jerusalem Municipality announced a plan to demolish 88 homes in Al-Bustan, which would have resulted in the displacement of some 1,500 residents (Ma'an, 2017). The plan was lobbied by the controversial settler organisation Ir-David Foundation which in the 1990s was mandated to manage the City of David's national park and lead archaeological excavations in the Holy Basin (Bronner, 2008; Ir-Amim, 2012). Subsequently, they started to pursue their self-proclaimed goal of taking control of Silwan, considered an important historical site, and strengthen its Jewish character (Freedman, 2008; Landy, 2017). After the residents' appeals to the relocation plan were rejected, the Jerusalem Municipality proposed that they voluntarily move to another Palestinian neighborhood of Beit Hanina, in northern East Jerusalem, but the residents refused (Ma'an, 2017).

In early 2010, the municipality unveiled a new plan for a 'King's Garden' in Al-Bustan, which designated the neighbourhood as a mixed tourism-housing area, whereby 22 residential properties would be evicted, while other parts would be developed into hotels, restaurants, art shops and a park area (Ma'an, 2017; Gadzo, 2017). Despite allegations by NGOs and residents that 56 buildings – rather than 22 – were slated for demolition, the plan was subsequently approved by the planning authorities (Ir-Amim, 2012).

Since then, Israeli forces have regularly raided the Al-Bustan neighbourhoods and issued various demolition orders to residents (Ma'an, 2017). According to Ir-Amim (2012), a range of strategies have been employed to transfer ownership of properties in Silwan, including:

- seizure of houses declared to be "absentee property" (based on a broad interpretation of the absentee property law generally opposed by attorneys in the past);
- the transfer of properties claimed to have been owned by Jews before 1948;
- purchase of properties in convoluted transactions; and
- the massive transfer of public properties and land to the exclusive control of Elad with no proper or transparent administrative process.

(pp. 14–15)

At the end of 2011, the planning authorities recommended additional building projects in Silwan involving the Ir David Foundation, namely a visitor centre, a museum and additional excavation sites (Ir-Amim, 2012). In 2015, the Ir David Foundation together with the Israel Nature and Parks Authority began to close off large areas of the archaeological park with fences and gates, making it impossible for Silwan residents to enter the area after 5pm and on weekends and holidays (Hasson, 2017). In November 2017, Israel's High Court of Justice ruled that the authorities and the Ir David Foundation must first find alternative areas for the Palestinian residents, before denying passage into and through the enclosed sites (ibid.). One month earlier, international media reported that eleven homes had already been destroyed to make way for the park since the first demolitions in 2005 (Gadzo, 2017).

Israeli authorities and the Ir David Foundation ensure that tourist experiences in the 'City of David' area are not 'disturbed' by Palestinian residents. While previously tour groups visiting the archaeological sites had to return to the Old City through the Wadi Hilweh Street (a Palestinian residential zone), Ir David built an alternative uphill walkway that passes through well-maintained settler Jewish homes (Landy, 2017). Underground tunnels are also excavated to guarantee a 'Palestinian-free' passage, which in several cases have shaken the foundations of Palestinian houses and caused some of them to collapse (Miller, 2013; Terrestrial Jerusalem, 2018).

The village of Silwan is not the only battleground on which Palestinian residents clash with expansionist tourism and settlement plans of Jewish Israelis. Sheikh Jarrah is another contentious Palestinian neighbourhood in East Jerusalem, which is culturally and touristically significant because of its proximity to the tomb of Simeon the Just, a Jewish High Priest during the time of the Second Temple. In the early 1970s, Israel's Custodian General re-allocated the properties to two Jewish trusts, thereby forcing the Palestinian residents who had earlier been promised ownership to pay rent to Israeli landlords instead (Miller, 2013).

Another contentious issue is the planned Jerusalem cable car project, a government initiative promoted by the Jerusalem Development Authority (JDA) and the Ministry of Tourism which declared "the project a 'national priority', a category usually reserved for advancing massive infrastructure and road construction projects" (Ezrahi and Mizrachi, 2020, p. 140). The project was approved by the Israeli Cabinet in early 2019, aiming to connect West Jerusalem with the Old City and the Mount of Olives by 2021 (Hayun, 2019; Kimmelman, 2019). The US$60 million project was innocuously presented as a 'green solution' to address problems of traffic congestion and accessibility but opponents have argued that it is another attempt to present a one-sided narrative of Jerusalem and advance Israeli ownership claims over Palestinian parts of the city (Kimmelman, 2019; Scammell, 2020). The planned route would pass over cemeteries, ancient churches and Palestinian homes in the Silwan neighbourhood and end on the rooftop of Ir David Foundation's planned visitor centre (Schonberg, 2019). In an early planning phase of the controversial project, then Mayor of Jerusalem Nir Barkat was quoted by Israeli media saying that the cable care will "bring the wider world to understand who really owns this city" (Hasson, 2016, n.p.). The project has not only drawn strong criticism from Palestinian advocacy groups, but also from Israeli environmentalists, archaeologists and urban planners who are concerned about the 'Disneyfication' of the city's cultural heritage (Kimmelman, 2019; Ezrahi and Mizrachi, 2020). The Israeli NGO Emek Shaveh joined forces with Silwan residents and submitted a petition to the Supreme Court, arguing that the hearings on the cable car plans had been flawed, the approval by an interim government had been illegitimate and the proponents of the project had downplayed "the serious harm it would cause to Jerusalem's cultural and historical heritage" (Schonberg, 2019, n.p.).

Hence, it seems obvious that the planning of archaeological parks and tourism facilities by the Israeli Government and the Jerusalem Municipality in collusion with right-wing settler organisations such as the Ir David Foundation does not primarily serve to protect people's rights to cultural heritage assets but pursues the broader political goal of limiting the development of the Palestinian population, appropriating their land and bolstering Israeli occupation. This goal is also reflected in the production of promotional tourism material. A critical and self-reflective exhibition 'Welcome to Jerusalem' in the Jewish Museum in Berlin, Germany, in December 2018 showed a former city map of Jerusalem that had been produced by Israel's Ministry of Tourism in 2016 (Figure 6.1). The 57 touristic sites depicted in the map include only one Muslim and five Christian places of interest. The area to the east of the Old City Wall is presented as an uninhabited, lush green hillside area, despite being a densely populated area, including the village of Silwan. The Israeli opposition criticised the map as a deliberate show of disrespect toward Palestinian, Muslim and Christian Jerusalem. A massive protest by churches and Palestinian organisations led to the publication of a revised edition (Figure 6.2).

In 2017, Israel's Minister of Tourism had to defend government plans to bar tour operators from taking visitors to Palestinian areas of the West Bank. The Jerusalem Post quoted him saying:

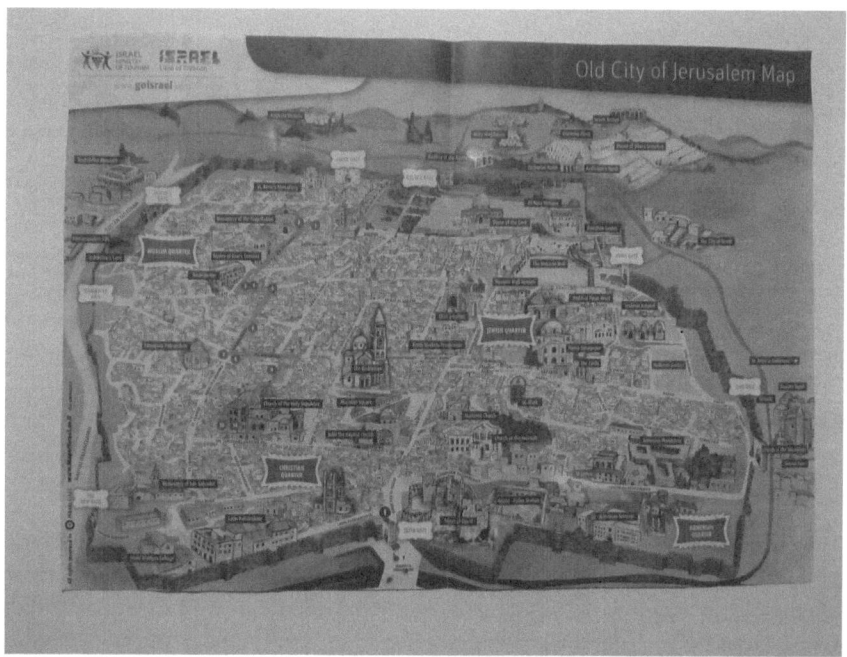

Figure 6.1 Map of Old City of Jerusalem (produced in 2016)

Figure 6.2 Map of Old City of Jerusalem (revised edition)

> We are not preventing anything. But, of course, we give incentives and
> encourage those who sell Israel, not those who sell other products ...
> We're responsible for bringing tourists to Israel, not to other places. And I
> do believe that the overall tourist experience for people staying in Israel is
> much better regarding every aspect, from nightlife to the quality of service
> to the personal security.
>
> (Schindler, 2017, n.p.)

In sum, the Israeli authorities not only annexed significant archaeological sites
and monuments and encroached on Palestinian cultural heritage in the name
of tourism, thereby depriving Palestinians of potential benefits from foreign
visitors (cf. Kassis, 2013). They have also instrumentalised international and
domestic tourism to erase Palestinian presence at these sites, while presenting
Israel as the sole guarantor of an authentic and secure visiting experience.

Concluding remarks

This chapter has demonstrated how under both authoritarian and democratic
rule, military and tourism can form alliances that control 'unruly' minority
groups, former enemies and occupied populations. Conflict and post-conflict
situations provide a fertile ground for the emergence of a military-tourism
complex (cf. Weaver, 2011) whereby armed forces play a pivotal role not only
in the securitisation of the tourism industry but also in its further expansion

both spatially and economically. In a post-conflict context, the military renders itself invisible behind ambitious tourism development plans and glossy marketing devices. Disguised as innocuous tourism actors and legitimised by a 'tourism and security' discourse (see Chapter 2), military forces instrumentalise the industry to seize land and displace communities, often under the radar of civil society. The durability of this military-tourism complex is a consequence of its adaptability between times of war and peace and the desire of the majority population to maintain the status quo and of the cosmopolitan tourist to enjoy a safe and secure vacation.

Securitisation is also one of the themes that run through the case studies of the next chapter which looks into the impacts of wildlife tourism and fortress conservation on involuntary relocation. The chapter will introduce protected areas as enclosed spaces of commodified nature for touristic consumption that are associated with various forms of dispossession and displacement. Examples will be drawn from Tanzania, Mozambique, India and Colombia.

References

Adnan, S. and Dastidar, R. (2010) *Alienation of the Lands of Indigenous Peoples in the Chittagong Hill Tracts of Bangladesh*. Chittagong Hill Tracts Commission and International Work Group for Indigenous Affairs: Dhaka, Copenhagen. Available at: www.iwgia.org/images/publications//0507_Alienation_of_the_lands_of_IPs_in_the _CHT_of_Bangladesh.pdf (accessed 12 January 2019).

Ahmed, H.S. (2017) *Tourism and State Violence in the Chittagong Hill Tracts of Bangladesh*. Master's thesis, The University of Western Ontario. Electronic Thesis and Dissertation Repository, 4840. https://ir.lib.uwo.ca/etd/4840t.

Alluri, R.M. (2009) *The Role of Tourism in Post-Conflict Peacebuilding in Rwanda*. Swisspeace: Bern.

Carter, C. (2015) '*Winners and losers: Land grabbing in the new Myanmar*' in Carter, C. and Harding, A. (eds) *Land Grabs in Asia: What Role for the Law?* Routledge: London, New York, pp. 100–117.

Causevic, S. and Lynch, P. (2011) 'Phoenix tourism: Post-conflict tourism role', *Annals of Tourism Research*, 38 (3): 780–800.

Chakma, B. (2018) 'Bangladesh' in IWGIA (ed.) *The Indigenous World in 2018*. International Work Group for Indigenous Affairs (IWGIA): Copenhagen, pp. 362–370.

Chakma, M. (2017) *Tourism development in the Chittagong Hill Tracts, Bangladesh: The impact on indigenous peoples (IPs)*. Master's thesis, Flinders University, Adelaide.

Chakma, M.K. and Chakma, S. (2015) Tourism and Development in Indigenous Territories: Partnership of Indigenous Peoples and Role of the Government. Presented at a roundtable organized by Kapaeeng Foundation20 August 2015 at CIRDAP Auditorium in Dhaka.

Clifton, J., Hampton, M.P. and Jeyacheya, J. (2018) 'Opening the box? Tourism planning and development in Myanmar: Capitalism, communities and change', *Asia Pacific Viewpoint*, 59 (3): 323–337.

Draper, R. (2016) 'Fragile peace', *National Geographic*, 230 (5): 108–129.

Ezrahi, T. and Mizrachi, Y. (2020) 'Cable Car Plan Threatens Unique Character and Heritage of the Old City of Jerusalem' in World Heritage Watch (ed.) *World Heritage*

Watch Report 2020. Berlin, pp. 138–142. Available at: https://world-heritage-watch. org/wp-content/uploads/2020/06/WHW-Report-2020.pdf (accessed 12 July 2020).

Gonzalez, V.V. (2013) *Securing Paradise: Tourism and Militarism in Hawai'i and the Philippines*. Duke University Press: Durham, London.

Henderson, J.C. (2003) 'The politics of tourism in Myanmar', *Current Issues in Tourism*, 6 (2): 97–118.

International Crisis Group (2019) *Fire and Ice: Conflict and Drugs in Myanmar's Shan State*. Available at: https://d2071andvip0wj.cloudfront.net/299-fire-and-ice.pdf (accessed 12 July 2020).

Ir-Amim (2012) *The Giant's Garden: The "King's Garden" Plan in Al-Bustan*. Research Report. Ir-Amim, Jerusalem.

Ir-Amim (2015) *Displaced in their Own City*. Research Report. Ir-Amim, Jerusalem.

IWGIA (2012) *Militarization in the Chittagong Hill Tracts, Bangladesh – The Slow Demise of the Region's Indigenous Peoples*. International Work Group for Indigenous Affairs (IWGIA): Copenhagen.

Kassis, R. (2013) *Tourism and Human Rights in Palestine*. Tourism Watch, Newsletter No. 72 (September 2013). Available at: www.tourism-watch.de/en/content/tourism -and-human-rights-palestine (accessed 17 December 2018).

Klein, N. (2007) *The Shock Doctrine: The Rise of Disaster Capitalism*. Penguin Group: London.

KPSN (2020) 'Gambling away our lands: Naypyidaw's "Battlefields to Casinos" strategy in Shwe Kokko'. Available at: https://progressivevoicemyanmar.org/wp -content/uploads/2020/03/Gambling-Away-Our-Lands-English.pdf (accessed 17 July 2020).

Kraak, A.-L. (2017) 'World heritage conservation and human rights in Bagan, Myanmar', *Historic Environment* 29 (3): 84–96.

Kramer, T. (2009) *Burma's Cease-Fires at Risk: Consequences of the Kokang Crisis for Peace and Democracy*. Transnational Institute: Amsterdam. Available at: www.tni.org/files/ download/psb1.pdf (accessed 17 July 2020).

Landy, D. (2017) 'The place of Palestinians in tourist and Zionist discourses in the 'City of David', occupied East Jerusalem', *Critical Discourse Studies*, 14 (3): 309–323.

Lisle, D. (2016) *Holidays in the Danger Zone: Entanglements of War and Tourism*. University of Minnesota Press: Minneapolis.

Mahr, J. and Sutcliffe, S. (1996) 'Come to Burma', *New Internationalist*, 280: 28–30.

Mohsin, A. (1997) *The Politics of Nationalism: The Case of the Chittagong Hill Tracts Bangladesh*. The University Press Limited: Dhaka.

Neef, A. (2016) *Land Rights Matter! Anchors to Reduce Land Grabbing, Dispossession and Displacement: A Comparative Study of Land Rights Systems in Southeast Asia and the Potential of National and International Legal Frameworks and Guidelines*. Bread for the World: Berlin.

Neef, A. and Grayman, J.H. (2018) 'Introducing the tourism-disaster-conflict nexus' in Neef, A. and Grayman, J.H. (eds) *The Tourism-Disaster-Conflict Nexus*. Emerald Publishing: Bingley, pp. 1–31.

Novelli, M., Morgan, N. and Nibigira, C. (2012) 'Tourism in a post-conflict situation of fragility', *Annals of Tourism Research*, 39 (3): 1446–1469.

Oberndorf, R. (2012) *Legal Review of Recently Enacted Farmland Law and Vacant, Fallow and Virgins Lands Management Law: Improving the Legal & Policy Frameworks Relating to Land Management in Myanmar*. Forest Trends Association: Washington DC. Available at: www.forest-trends.org/wp-content/uploads/imported/fswg_lcg_legal-review-of-

farmland-law-and-vacant-fallow-and-virgin-land-management-law-nov-2012-eng-2
-pdf.pdf (accessed 12 March 2016).

Parnwell, M.J.G. (1998) 'Tourism, globalization and critical security in Myanmar and Thailand', *Singapore Journal of Tropical Geography*, 19 (2): 212–231.

Pilger, J. (1996) 'The Burmese gulag', *CovertAction Quarterly*, 58: 6–15.

Piyadasa, T. (2016) *Development by Dispossession? Forced Evictions and Land Seizures in Paanama, Sri Lanka*. Oxfam International and People's Alliance for Rights to Land (PARL): Oxford.

Ratnayake, I. and Hapugoda, M. (2017) 'Land and tourism in post-war Sri Lanka: A critique on the political negligence in tourism' in Saufi, A., Andilolo, I.R., Othman, N. and Lew, A.A. (eds) *Balancing Development and Sustainability in Tourism Destinations*. Springer: Singapore, pp. 221–231.

Rhoads, E. (2018) 'Forced evictions as urban planning? Traces of colonial land control practices in Yangoon, Myanmar', *State Crime*, 7 (2): 278–305.

Seekins, D. (2011) *State and Society in Modern Rangoon*. Routledge: London, New York.

Seoighe, R. (2016) 'Inscribing the victors land: Nationalistic authorship in Sri Lanka's postwar Northeast', *Conflict, Security and Development*, 16 (5): 443–471.

Scurrah, N., Hirsch, P. and Woods, K. (2015) *The Political Economy of Land Governance in Myanmar*. Mekong Region Land Governance: Vientiane.

Sims, K. (2017) 'Gambling on the future: Casino enclaves, development, and poverty alleviation in Laos', *Pacific Affairs*, 90 (4): 675–699.

Sri Lanka Campaign for Peace & Justice (2018) 'Whose paradise? A guide to ethical tourism in Sri Lanka'. Available at: www.srilankacampaign.org/wp-content/uploads/2018/10/A-Guide-to-Ethical-Tourism-in-Sri-Lanka-October-2018-Compressed.pdf (accessed 12 November 2018).

STP (2016) *Under the Military's Shadow: Local Communities and Militarization on the Jaffna Peninsula*. Society for Threatened Peoples (STP): Ostermundigen.

Than, T. (2015) 'Black territory to land of 'paradise': The changing political and social landscape of Mongla' in Chan, Y.W. (ed.) *The Age of Asian Migration*, vol. 2. Cambridge Scholars Publishing: Newcastle upon Tyne, pp. 131–156.

Than, T. (2016) 'Mongla and the borderland politics of Myanmar', *Asian Anthropology*, 15 (2): 152–168.

Transnational Institute (2013) 'Access denied: Land rights and ethnic conflict in Burma'. Burma Policy Briefing. Available at: www.tni.org/files/download/accesde nied-briefing11.pdf (accessed 12 July 2020).

Weaver, A. (2011) 'Tourism and the military: Pleasure and the war economy', *Annals of Tourism Research*, 38 (2): 672–689.

Yokota, Y. (1996) *Report on the Situation of Human Rights in Myanmar*. UN Commission on Human Rights: Geneva. Available at: https://digitallibrary.un.org/record/228428 (accessed 12 July 2020).

Zhou, P. (2005) 'Troubling travels: Funding Myanmar's junta', *Harvard International Review*, 26 (4): 9–10.

Media sources/websites

Aye, M.N. (2015) 'Tada-U developers start on hotel zone', *Myanmar Times*, 19 January. Available at: https://mmtimes.com/business/property-news/12839-tada-u-develop ers-start-on-hotel-zone.html (accessed 12 July 2020).

Barua, S.K. (2020) 'New five-star hotel 'threatens' ancestral land of Mro community in Bandarban', *The Daily Star*, 7 October. Available at: www.thedailystar.net/country/news/mro-people-urge-pm-save-their-ancestral-villages-bandarban-1974101 (accessed 25 October 2020).

Bronner, Y. (2008) 'Archaeologists for hire: A Jewish settler organisation is using archaeology to further its political agenda and oust Palestinians from their homes', *The Guardian*, 1 May. Available at: www.theguardian.com/commentisfree/2008/may/01/archaeologistsforhire (accessed 1 January 2019).

Chakma, N. (2017) 'Bengali tourists in CHT', *New Age*, 12 July. Available at: www.newagebd.net/article/17553/bengali-tourists-in-cht (accessed 12 January 2019).

Chau, T. and Htet, K. (2019) 'Bagan named UNESCO World Heritage Site', *Myanmar Times*, 7 July. www.mmtimes.com/news/bagan-named-unesco-world-heritage-site.html (accessed 12 July 2020).

Freedman, S. (2008) 'Digging into trouble'. *The Guardian*, 26 February. Available at: www.theguardian.com/commentisfree/2008/feb/26/diggingintotrouble (accessed 2 January 2019).

Gadzo, M. (2017) 'Silwan demolitions: "They're destroying Jerusalem"', *Al Jazeera*, 9 October. Available at: www.aljazeera.com/indepth/features/2017/09/silwan-demolitions-destroying-jerusalem-170920080554388.html (accessed 2 January 2019).

Gadzo, M. (2018) 'Palestinians battle home evictions in East Jerusalem's Silwan', *Al Jazeera*, 8 December. Available at: www.aljazeera.com/indepth/features/palestinians-battle-home-evictions-east-jerusalem-silwan-181206052617178.html (accessed 2 January 2019).

Hasson, N. (2016) 'Jerusalem Mayor: Cable car stop in Palestinian neighborhood will clarify "who really owns this city"', *Hareetz Newspaper*, 25 August. Available at: www.haaretz.com/israel-news/.premium-barkat-east-jerusalem-cable-car-will-clarify-who-really-owns-city-1.5428939 (accessed 5 July 2020).

Hasson, N. (2017) 'Right wing Jewish organization ordered to develop open space for Palestinians', *Hareetz Newspaper*, 27 November. Available at: www.haaretz.com/israel-news/.premium-right-wing-jewish-organization-ordered-to-develop-open-space-for-palestinians-1.5626799 (accessed 2 January 2019).

Hayun, D. (2019) 'Israel's planned Jerusalem cable car irks Palestinians', *Reuters*, 8 November. Available at: www.reuters.com/article/us-israel-palestinians-cable-car/israels-planned-jerusalem-cable-car-irks-palestinians-idUSKBN1XH23A (accessed 5 July 2020).

Hurng, H. (2020) 'Shan community wants palace land returned', *Shan Herald Agency for News*, 7 February. Available at: www.bnionline.net/en/news/shan-community-wants-palace-land-returned (accessed 12 July 2020).

International Tribunal on Evictions (2017) 'Case from Sri Lanka: Land grabbing for tourism development, Panaama Village', Available at: www.tribunal-evictions.org/international_tribunal_on_evictions/evictions_cases/6th_session_2017/case_from_sri_lanka_land_grabbing_for_tourism_development_panaama_village (accessed 12 January 2019).

Kapaeeng Foundation (2018) Government plans to acquire 700 acres of land for establishment of Special Tourism Zone at Alutila hills in Khagrachari which leads to eviction of hundreds indigenous families. Available at: www.kapaeng.org/government-plans-to-acquire-700-acres-of-land-for-establishment-of-special-tourism-zone-at-alutila-hills-in-khagrachari-which-leads-to-eviction-of-hundreds-indigenous-families/ (accessed 12 January 2019).

Kimmelman, M. (2019) 'Cable cars over Jerusalem? Some see "Disneyfication" of Holy City', *New York Times*, 16 September. Available at: www.nytimes.com/2019/09/13/world/middleeast/jerusalem-cable-cars.html (accessed 5 July 2020).

Laskin, D. (2013) 'Shake up at City of David', *Jerusalem Post*, 13 April. Available at: www.jpost.com/Features/In-Thespotlight/Shake-up-at-City-of-David-309761 (accessed 3 January 2019).

Lonely Planet (2020) 'Sri Lanka', Available at: www.lonelyplanet.com/sri-lanka (accessed 12 June 2020).

Lwin, N. (2020) 'Myanmar govt to probe contentious Chinese development on Thai border', *The Irrawaddy*, 16 June. Available at: www.irrawaddy.com/news/burma/myanmar-govt-probe-contentious-chinese-development-thai-border.html (accessed 12 July 2020).

Ma'an (2017) 'Israeli authorities deliver 16 demolition orders in Silwan', 11 February. Available at: www.maannews.com/Content.aspx?id=775407 (accessed 3 January 2019).

Miller, A.L. (2013) 'Israel's land grab in East Jerusalem', *The Nation*, 17 April. Available at: www.thenation.com/article/israels-land-grab-east-jerusalem/ (accessed 1 January 2019).

New Age (2016) Nat'l minorities want project Alutila tourism zone scrapped. *New Age*, 4 September. Available at: http://archive.newagebd.net/250028/natl-minorities-want-project-alutila-tourism-zone-scrapped/ (accessed 12 January 2019).

Nyein, N. (2019) 'Chinese developer's grand claims spark fresh concern in Karen state', *The Irrawaddy*, 6 March. Available at: www.irrawaddy.com/news/burma/chinese-developers-grand-claims-spark-fresh-concern-karen-state.html (accessed 12 July 2020).

Scammell, R. (2020) 'Jerusalem's "green" cable car plan is challenged in Israel's top court', *The World*, 25 February. Available at: www.pri.org/stories/2020-02-25/jerusalems-green-cable-car-plan-challenged-israels-top-court (accessed 5 July 2020).

Schindler, M. (2017) 'Tourism minister: Israel not restricting visitors to Palestinian areas', *Jerusalem Post*, 2 October. Available at: www.jpost.com/Arab-Israeli-Conflict/Israel-not-restricting-visitors-to-Palestinian-areas-says-tourism-minister-506590 (Accessed 3 January 2019).

Schonberg, D. (2019) 'The Jerusalem cable car', 17 December. Available at: www.jpost.com/opinion/the-jerusalem-cable-car-610932 (accessed 5 July 2020).

Soe, T.N. and Ko, K.K. (2014) 'Tade Oo hotel zone brings sleepless nights', *Myanmar Times*, 15 May. Available at: https://mmtimes.com/national-news/10361-tada-oo-hotel-zone-brings-sleepless-nights.html (accessed 12 July 2020).

Terrestrial Jerusalem (2018) 'The Israeli Government's creation of a settler realm in and around Jerusalem's Old City', 23 March. Available at: http://t-j.org.il/LatestDevelopments/tabid/1370/currentpage/1/articleID/876/Default.aspx (accessed 3 January 2019).

Tha, K.P. (2019) 'Despite world heritage status, Bagan's future far from assured', *The Irrawaddy*, 2 August. Available at: www.irrawaddy.com/features/despite-world-heritage-status-bagans-future-far-assured.html (accessed 12 July 2020).

The Daily Star (2016) 'Govt cancels Khagrachhari tourism project', *The Daily Star*, 6 October. Available at: www.thedailystar.net/backpage/govt-cancels-khagrachhari-tourism-project-1294657 (accessed 12 January 2019).

Thitsar, S.G. (2013) 'Seven charged over Inle Lake hotel zone protest', *Myanmar Times*, 25 February 2013. Available at: https://mmtimes.com/national-news/4195-seven-charged-over-hotel-zone.protest.html (accessed 12 July 2020).

Tripura, J. (2016) 'Tourism development: Triggered to destroy the life and culture of indigenous peoples'. Available at: https://johntripura.wordpress.com/2016/09/05/tourism-development-triggered-to-destroy-the-life-and-culture-of-indigenous-peoples/ (accessed 12 January 2019).

World Bank (2016) 'World Bank provides Bangladesh $360 million for improving inland waterways', *World Bank Press Release*, 16 June. Available at: www.worldbank.org/en/news/press-release/2016/06/16/provides-bangladesh-360-million-improving-inland-waterways (accessed 12 January 2019).

7 Wildlife tourism, fortress conservation and green grabbing

Over the past 150 years, more than 100,000 parks, or protected areas, have been established globally, covering between 13 and 15 per cent of the land surface of our planet. Commonly subsumed under the term "protected areas" are "all national parks, game reserves, national monuments, forest reserves and the myriad other places and spaces for which states provide special protection from human interference" (Brockington, Duffy and Igoe, 2008, p. 1). These parks have created leisure opportunities for hundreds of millions of tourists annually. Different forms of ecotourism, adventure tourism and wildlife tourism have been proposed to serve as a central source of revenue to maintain protected areas and sustain wildlife conservation efforts. Yet most tourists seem completely unaware of the paradox of contained 'wilderness', managed nature, and wild animals subjected to human rules.

Historically, protected areas have often been associated with the large-scale removal of the original human population to create 'people-free' zones and sustain 'wild' flora and fauna for exclusive, 'non-consumptive' use by privileged travellers. It has been estimated that about 130 million conservation refugees have lost their homes and livelihoods to parks (Survival International, n.d.), yet precise figures are not available as most countries do not keep displacement records and the literature on conservation-induced displacement remains "patchy and parlous" (Brockington and Igoe, 2006, p. 452). Apart from physical displacement, protected areas have also contributed to economic displacement by denying or restricting the original inhabitants access to customary livelihood sources, such as timber, water, meat, non-timber forest products and other essential resources. Brockington, Duffy and Igoe (2008, p. 71) also describe how the delineation of parks "displace people symbolically" by writing "them out of landscape's history, proclaiming that they do not belong".

The subsequent section provides a brief overview of how evictions of Indigenous peoples from protected areas have become regularised events since colonial times. This is followed by case studies on various forms of displacement from national parks and wildlife sanctuaries in Eastern and Southern Africa as well as in India and Colombia.

Protected areas and displacement of Indigenous and forest-dependent peoples: a brief historic perspective

Brockington, Duffy and Igoe (2008) contend that the history of protected areas dates back hundreds if not thousands of years, citing examples from ancient empires in today's China, India, Indonesia, Syria, Iran, Iraq and Mongolia. Notwithstanding the pre-colonial history of protected area development, the Yellowstone National Park in the United States is widely considered the world's first national park, established in 1872. Its creation cost the lives and livelihoods of many members of the Shoshone, Bannock, Blackfoot and Crow peoples who had sustainably lived in the Yellowstone region for thousands of years before the Europeans arrived in the 'New World' (Quammen, 2016; Survival International, n.d.). This model of forced displacement and exclusion of Indigenous and forest-dependent peoples for conservation and 'nature tourism' has since been exported worldwide, most often with devastating impacts (Survival International, n.d.). Colonial displacements from protected areas have often been driven by an imaginary 'wilderness' (Akama, 2004; Lunstrum and Ybarra, 2018). Other settler states, such as New Zealand, Australia and Canada, swiftly followed the US national park model, and many postcolonial governments adopted the 'fortress conservation' approach and continued to evict Indigenous peoples from 'ecologically sensitive' areas, often on the grounds that such measures contribute to national and local development, preserve the country's natural heritage and lead to improved security, as many protected areas lie in remote and politically sensitive border regions.

Protected areas have often played a major role in state territorialisation and nation-building (Forsyth and Walker, 2008; Brockington, Duffy and Igoe, 2008). By removing 'messy' customary rights of Indigenous peoples through the imposition of a uniform state property regime, many nation states have systematically extended and consolidated their power over large tracts of land. Opening these protected areas to domestic visitors and foreign tourism provides additional legitimacy for demarcating and securitising the areas and removing any former residents that are considered as 'eco-threats'. Most of the evictions from protected areas are highly racialised, as some of the examples in the following sub-sections will show. According to Brockington, Duffy and Igoe (2008), evictions have been most common in Africa, South Asia and Southeast Asia, while much fewer cases are reported from South America, the Caribbean and the Pacific. In sub-Saharan Africa, reports about conservation-related human rights abuses – evictions, beatings, burning of houses – of Indigenous peoples, such as the Baka Indigenous people in Cameroon, the Maasai pastoralists in Kenya and Tanzania (see the following section), and the Kalahari Bushmen in Botswana recently made headlines in the international media and have been the subject of various academic studies (Vidal, 2016).

In many parts of Latin America, conflicts over conservation areas and tourism have pitched local communities against national governments and private sector interests. In Los Esteros del Iberá, Argentina's largest nature reserve, for instance, a private US American conservation enterprise acquired 150,000 hectares of old cattle ranches for conservation and tourism purposes, which raised questions among the Indigenous communities as to why a foreigner was able to obtain such a huge amount of land in a sovereign state (Busscher, Parra and Vanclay, 2018). While some activists denounced this large-scale land acquisition as a neo-colonial green grab and suspected a hidden agenda (such as water grabbing or the start of a military base), other groups showed admiration for the ostensibly philanthropic intentions of the project (Busscher, Parra and Vanclay, 2018).

The establishment of marine protected areas (MPAs) started in the early 20th century and has lagged behind terrestrial protected areas in terms of the timing, extent and effectiveness of their creation and coverage. In recent years, the concept of ocean and marine grabbing has arisen, which includes cases where the delineation of marine parks or other protected areas in coastal and ocean environments has led to the dispossession of previous customary users. McClanahan and Mangi (2000) identified a 60–80 per cent decline in the number of fisherfolks after the establishment of a no-take Marine Park at the Jomo Kenyatta Beach in Mombasa, Kenya's largest and most popular public beach. In Malaysia's Redang Island Marine Park, Hill (2017) found that the livelihoods of fishers became increasingly compromised by a no-take zone stretching two nautical miles from the coastline, which was established to promote nature-based tourism. Benjaminsen and Bryceson (2012) have described how marine conservation in the Mafia Island Marine Park in Tanzania dispossessed local inhabitants and shifted access to and control over marine resources from fishers and other customary users to more powerful actors, such as government officials, tourism operators and transnational conservation organisations.

Tourism-driven evictions from state and privately managed protected areas in Eastern and Southern Africa

In Eastern and Southern Africa, the designation of wildlife reserves for photo safari and hunting tourism goes back to the establishment of colonial rule and has displaced thousands of local communities from customary land on which they relied for their livelihoods. Safari hunting was a favourite pastime for colonial administrators, and the period between 1900 and 1945 in East Africa has often been referred to as the 'Era of Big Game Hunting' (Akama, 2004). In South Africa, game reserves and national parks, such as the Kruger National Park, became "manifestations of 'apartheid repression'" and part of "a process of alienating Africans from wildlife" (Carruthers, 1995, pp. 89–90), often associated with forced evictions of Indigenous communities.

Safari tourism and hunting concessions in Tanzania

Land grabbing and land conflicts have a long tradition in Tanzania and most often involve pastoralist groups. After Germany established colonial rule over what they called 'German East Africa' following the Berlin Conference of 1884–1885, Maasai pastoralists were evicted from their traditional grazing grounds in West Kilimanjaro onto barren plains, as the Germans grabbed the most productive land for the establishment of plantations and hunting concessions. When the British took over as colonial power during World War I, they occupied the fertile land in the temperate slopes of Mount Kilimanjaro, hence the Maasai lost even more of their traditional grazing grounds (Porokwa, 2018). When colonial powers started to ponder over independence of their African colonies, European conservationists warned that Africans could not be trusted taking care of wildlife (Pearce, 2012). One of the strongest and most controversial messages came from German zoo director and animal conservationist Bernhard Grzimek in his influential book and film *Serengeti Shall Not Die* where he stated that "[a] National Park ... must remain a primordial wilderness to be effective. No men, not even native ones, should live inside its borders" (cited in Adams and McShane, 1992, p. xvi).

Following Tanzania's independence, evictions continued under various governments (Ngoitiko et al., 2010). The dramatic landscape of Ngorongoro in Tanzania – roughly equally divided between the Serengeti National Park and the Ngorongoro Conservation Area – is one of the world's most famous conservation areas. Few visitors realise, however, that in the 1970s, evictions

Figure 7.1 A group of giraffes in a Tanzanian wildlife reserve

Figure 7.2 A Maasai takes his livestock to a water hole

from the Serengeti National Park pushed thousands of Maasai and their animals into the Ngorongoro Crater, where they were not allowed to graze their animals and from where many were subsequently evicted again (Mowforth and Munt, 2016). The famous crater has now become severely degraded, and UNESCO has threatened to remove its World Heritage status. In early 2010, the government responded by calling for the removal of the thousands of Maasai who live in the crater (Survival International, n.d.). In 2013, the then Tanzanian Minister of Tourism proclaimed the establishment of a 1,500km^2 'wildlife corridor', ostensibly to allow wildlife to move freely between the Serengeti National Park and the Maasai Mara National Park in Kenya. This decision posed a severe threat to the livelihoods of several thousands of Maasai (Mowforth and Munt, 2016).

One of the most notorious conflicts between trophy hunting tourism and pastoralism has been going on for nearly 25 years in Loliondo, an area in proximity of the Serengeti National Park. In 1992 an investor from the United Arab Emirates obtained hunting rights over an area inhabited by several Maasai communities that had been in possession of documented land ownership rights (Benjaminsen et al., 2013; Massé and Lunstrum, 2016; for a brief discussion on trophy hunting, see Box 7.1). The Maasai strongly resisted the hunting concession and – to show their disapproval – signed contracts with private tourism companies that organised photo safaris for foreign tourists (Gardner, 2012).

Box 7.1 Should we kill wild animals to save them? The controversy around trophy hunting

Trophy hunting – the killing of large animals for their horns, tusks, skins or taxidermied bodies – has become a multi-billion-dollar industry. Several countries in sub-Saharan Africa allow this form of extractive tourism, with varying degrees of transparency and accountability. Some countries have established quotas or created exclusions to reflect the status of vulnerable game populations. For instance, South Africa no longer allows the hunting of leopards, while Namibia banned the use of dogs for hunting leopards after numbers were falling drastically. Botswana declared a temporary ban on trophy hunting in government-controlled hunting areas in 2014. The Kenyan government has completely banned trophy hunting since 1977.

Proponents of trophy hunting argue that the industry makes significant contributions to the economy of a number of African countries and provide much-needed funds for conservation and anti-poaching efforts. Some African governments contend that people in other continents have no right to tell them how they should manage their own wildlife and likened any involvement by Western environmental activists to neocolonialism. Critics of the practice maintain that no country should be involved in the business of selling and profiting from dead wildlife and argue that most of the profits made from trophy hunting do not go back into conservation and anti-poaching schemes but rather feed endemic corruption in host countries. They call for more financial contributions to wildlife protection from multi-lateral development banks, eco-philanthropists and NGOs.

Sources: Lindsey et al., 2012; Paterniti, 2017

The collaboration of the Maasai with private tour operators in Loliondo was a thorn in the side of the Tanzanian government for a long time. The government formally holds the ownership and control over the country's wildlife, even if the land itself is controlled by communities through their village councils (Nelson and Blomley, 2010; Massé and Lunstrum, 2016). The conflict escalated in 2009 when the government evicted about 200 Maasai families from the area, burned their huts and killed thousands of livestock. The eviction – while being a gross violation of human rights law – was made legally possible by the Wildlife Conservation Act of 2009, which prohibits grazing animals in so-called Game Controlled Areas (Benjaminsen and Bryceson, 2012). The Tanzanian government justified the forced relocation of the Maasai by the discourses of 'national development' and 'protection of wildlife' whose seasonal migration was deemed to be threatened by the pastoralist herds (Gardner, 2012). The violent relocation of the Maasai triggered national and international condemnation and risked damaging the Tanzanian government's and the UAE-based investor's reputation. Consequently, the actors involved tried to establish a more peaceful relationship with the Maasai and for several years, the conflict seemed to have abated.

Despite the short period of peace, in August 2017, Loliondo was in domestic and international headlines again, when the Ngorongoro District Commissioner issued an order to evict Maasai from their legally registered village lands in the so-called buffer zone of the Serengeti National Park. The eviction order was preceded by various statements of the then Minister of Natural Resources and Tourism, who called for the alienation of at least 1,500km^2 of Loliondo (about 40 per cent of the entire Loliondo area – roughly the size of London), which had been occupied by the Maasai community for centuries. Authorities were given an ultimatum of five days for carrying out the order. The evictions affected about 350 Maasai families who saw their *boma* (traditional houses) burnt to the ground and most of their property destroyed, leaving them without shelter, food or water (Porokwa, 2018). In September 2017, the Loliondo Maasai filed a legal case at the East African Court of Justice against the Tanzanian government (IWGIA, 2019). Following pressure from civil society organisations and international advocacy groups, the Tanzanian government halted the evictions in November 2017, followed by a statement from the new Minister of Natural Resources and Tourism that they had been illegally designed, without following proper procedures. Yet, the future of the remaining Loliondo Maasai still hangs in the balance, and in late 2018 and early 2019, two Maasai human rights activists were repeatedly detained without bail for alleged sedition (IWGIA, 2019b).

Meanwhile, across Tanzania, the Ministry of Natural Resources and Tourism has established about 20 wildlife management areas (WMAs) comprising around 30,000km^2 of land. This involved the technical and financial support of international donors, including the United States Agency for International Development (USAID), the German Agency for International Cooperation (GIZ), the German Development Bank (KfW) and the United Nations Development Programme (UNDP) (Bluwstein et al., 2018). The basic idea behind the WMA initiative is that communal land is set aside across several neighbouring villages to support the movement of wildlife between protected areas. The WMAs include bans on farming and settlements, and such activities as livestock grazing, firewood collection and charcoal making are often heavily restricted. In return, villagers are promised benefits from safari tourism revenues by cooperating with tourism investors (Bluwstein et al., 2018).

While the WMA model adopts the language of 'participation' and 'community-based conservation' and appears to be a triple-win arrangement for (1) wildlife conservation, (2) new livelihood opportunities for local communities, and (3) tourism businesses, in many instances the implementation of WMAs has triggered conflicts over land ownership and village boundaries and led to dispossession and evictions (Benjaminsen and Bryceson, 2012). Female-headed households who depend most on communal land resources for firewood and non-timber forest products have proven to be particularly disadvantaged by WMA establishment (Bluwstein et al., 2018). Maasai pastoralists have also not benefitted from WMA establishment, as at least half of their household income is derived from livestock compared to conservation-related tourism which accounts for less than five per cent (Burgoyne and Mearns, 2017). In the early days of the WMA programme, villagers had been promised hunting quotas and local control over trophy hunting, but the hunting industry proved highly lucrative for the Tanzanian government and remained under firm state control (Benjaminsen and Bryceson, 2012). In September 2018, 18 huts of an Indigenous Barabaig pastoralist community were burnt to the ground to make way for a tourism company that operates a facility in the Burunge WMA (IWGIA, 2019b).

Recently, a silver lining has appeared in the form of a protest movement driven by over 800 pastoralist women who successfully lobbied the Tanzanian leadership to resolve the land rights violations of pastoralist communities (IWGIA, 2019a). In 2019, the Tanzanian President spoke for the first time publicly against the expropriation of pastoralists' land in the name of wildlife conservation (IWGIA, 2019b).

Transnational sovereignty grabbing and evictions from the Great Limpopo Transfrontier Park for wildlife conservation and tourism, Mozambique and South Africa

The ongoing establishment of the Great Limpopo Transfrontier Park – a transboundary conservation initiative between South Africa, Zimbabwe and Mozambique – has been hailed by many environmentalists as a bold and innovative model of protecting wildlife across international borders (see Box 7.2). The megapark has been marketed as "a kingdom of animals like no other in the world" by South Africa's Peace Parks Foundation, a major driving force behind the park's establishment, alongside the African Wildlife Foundation. Only two days after the signing of the international treaty in December 2002, part of the boundary fences that separated the three countries was removed, and wild animals – such as elephants and other large mammals – were subsequently transferred from South Africa's Kruger National Park (KNP) into the Limpopo National Park (LNP) (Rodgers, 2009). Suddenly, the customary land rights of communities living within the LNP were in limbo.

Box 7.2 The creation of the Great Limpopo Transfrontier Park

In the late 1990s, South African conservation officials began to meet with counterparts in Mozambique and Zimbabwe to discuss possibilities for joining South Africa's Kruger National Park with adjacent parks and lands in neighbouring countries. After successful deliberations, the tri-country Great Limpopo Transfrontier Park (GLTP) was established in 2002/03. While Zimbabwe had a pre-existing park (the Gonarezhou National Park) and two other conservation areas to bring to the table, post-conflict Mozambique had no formally protected areas to contribute. Hence, with strong pressure from South Africa, Mozambique transformed a defunct colonial hunting reserve – known as 'Coutada 16' – into the Limpopo National Park (LNP). The German Development Bank (*Kreditanstalt für Wiederaufbau* – KfW) provided the initial funding to the Mozambican Ministry of Tourism to instigate the creation of the LNP and support the process of 'voluntary resettlement' of local communities from the park.

Before the LNP's designation in the early 2000s, Mozambican officials had proposed that the reserve should become a multi-functional zone that would allow co-existence between wildlife and humans. The area of about one million hectares is inhabited by more than 31,000 people. South African state and environmental NGO officials, however, insisted that the core zone be declared a more restrictive protected area requiring the removal of resident communities. Their argument was that inhabitants would disrupt the feel of wilderness that could bring tourism revenues and that they would threaten wildlife in the Kruger National Park (whose last remaining residents had been forcefully removed under the apartheid regime in the late 1960s). After South African demands prevailed, Mozambican officials also began to justify the relocation of communities from the park on the grounds that this was for their own safety, as the LNP was being restocked with 'dangerous' wildlife from the Kruger National Park.

Source: Lunstrum and Ybarra, 2018

A large part of the original inhabitants of the area covered by the LNP had already been displaced by Mozambique's long civil war from 1977–1992, yet there remained a significant number of local settlements, particularly concentrated within those areas that promised the highest potential for tourism and wildlife protection (Rodgers, 2009). When residents heard about the establishment of the new park, some reacted with threats of violent action against wildlife and tourists. Mozambican park officials assured them that they would not be forced to relocate from LNP, as forced displacement was unacceptable to the major donors. Funding from the German Development Bank (KfW) was contingent upon the relocation process being entirely voluntary.

Yet villagers' main livelihood activities – farming, livestock rearing and hunting – were considered incompatible with the LNP's conservation objectives. Therefore, park officials had to look for more subtle ways to make people move out of the park. The strategy of choice was to gradually tighten park regulations in order to limit villagers' livelihood options and also remind villagers that they would have much better livelihood opportunities elsewhere and would be safe from dangerous wildlife if they moved (DeMotts, 2017). While many park residents were in fact concerned about the dangers posed by roaming elephants and other large wildlife, most of them did not trust government officials due to prior experiences with forced displacement, hence the resettlement process was delayed by several years.

In 2003, the LNP management commissioned a so-called 'Resettlement Policy Framework' which drew on international best practice, i.e. the World Bank's safeguard policy on involuntary resettlement, which includes such principles as (1) avoid or minimise involuntary resettlement, (2) pursue resettlement as a development project, (3) engage in extensive consultation with communities, and (4) establish fair and rigorous grievance mechanisms (Rodgers, 2009). Two communities were eventually resettled – one in 2009 and the other in 2011 – following a series of consultations and after giving informed consent as a whole community, although there had been individual voices of dissent (Otsuki, Achá and Wijnhoud, 2017). The resettlement action plans had foreseen adequate compensation for all material losses, but did not take into account less tangible losses, such as leaving graveyards behind or losing access to traditional plants and animals that were only found in the core zone of the park (Witter and Satterfield, 2014).

While the agreement promised the provision of better housing, infrastructure and farmland – which were the main reasons why the community gave their consent – one of the communities had not obtained farmland nearly five years after the relocation and had to negotiate land access with a neighbouring community (Otsuki, Achá and Wijnhoud, 2017). The LNP management insisted that it was up to the respective district to provide farmland for the resettled people, while local government officials asserted that it was the community's own responsibility to make requests for land allocation to neighbouring communities. As a consequence, there was rising resentment within the community against the park officials and the local government (Otsuki, Achá and Wijnhoud, 2017). This case begs the questions of whether one can speak of 'voluntary' resettlement in a situation of limited choice and false promises.

Tiger conservation and tourism in India: precarious livelihoods at the human–wildlife interface

It is estimated that about one million people in India – many of them Indigenous and tribal peoples – live across 657 protected areas, composed of 99 national parks, 513 wildlife sanctuaries, 41 conservation reserves and four community reserves, which cover about five per cent of the total land mass of the world's second most populated country (Mahapatra, Tewari and Baboo, 2015). Relocation of forest-dependent communities out of protected areas, particularly from national parks and wildlife sanctuaries, is a recurring issue although it rarely makes it into the international headlines. A conservative estimate suggests that at least 100,000 people have been displaced from protected areas in India from the 1970s up to 2008 (Lascorgeix and Kothari, 2009).

There are about 50 tiger reserves in India visited by nearly 1.5 million people in 2014/2015, accounting for 32 per cent of all wildlife visits in the country (Ayyar, 2018). Consequently, there is an ongoing construction boom of hotels, resorts and eco-lodges along the boundaries of tiger reserves helping to fund conservation efforts but also raising concerns about encroachment of tourists on wildlife habitats. Historically, conservationists and authorities in India have been much more preoccupied with the ostensibly negative impact of original settlers on tiger habitats and have called for their removal from the core zone of tiger reserves. Eighty villages and 2,900 families have officially been relocated from tiger reserves since 1973, but the actual numbers are likely to be far higher (Hussain, Dasgupta and Singh Bargali, 2016).

Prior to 2006, the relocation of Indigenous and tribal peoples from tiger reserves was done without a clear legal framework, was rarely based on participatory processes and free, prior and informed consent (FPIC) and involved poor compensation packages (Shahabuddin and Bhamidipati, 2014). Yet in the mid-2000s, the Indian government introduced a number of changes to its national legislation related to wildlife conservation and resettlement, most notably the promulgation of the Forest Rights Act (FRA) (see Box 7.3). Since then a number of studies have been conducted to examine whether the new laws and policies have led to better outcomes for communities relocated from tiger reserves.

Box 7.3 India's wildlife conservation and resettlement laws and policies: A brief overview

Until the late 1990s, India had no specific policies and legal frameworks for conservation-induced displacement. Hence, communities living within tiger reserves and wildlife sanctuaries were considered unlawful residents and were liable to eviction. Resettlement processes were managed according to the Beneficiary-Oriented Tribal Development (BOTD) scheme, which consisted of a standardised compensation package of a two-hectare piece of land and INR 100,000 (about US$2,200) for land development, building material, community facilities and some other livelihood support measures. This proved insufficient for rehabilitating livelihoods in the new site, hence nearly all of the earlier resettlement schemes led to the socio-economic deterioration of relocated households.

In late 2006, after years of indigenous activism, the historic *Scheduled Tribes and Other Traditional Forest Dwellers (Recognition of Forest Rights) Act* (FRA) came into force. The FRA treats relocation for the cause of wildlife conservation as a last resort, to be carried out only after possible human-wildlife coexistence options have been considered. The FRA also requires that resettlement is conducted "on the basis of scientific criteria" and "after complete and just settlement of rights". The FRA coincided with substantive amendments to India's Wildlife Protection Act (WLPA), e.g. the inclusion of Section 38 which states, "No resettlement shall take place until facilities and land allocation at the resettlement location is complete as per the promised package." A revised Centrally Sponsored Scheme for Integrated Development of Wildlife Habitats prescribes a wider consultative process and free, prior and informed consent by village councils and individual families before relocation. Another significant improvement is a ten-fold increase of the compensation package to INR 1 million (about US $22,200), while the entitlement to the amount of land (2ha) remained unchanged.

Sources: Shahabuddin and Bhamidipati, 2014; Dash and Bahera, 2018

A comparative study of the quality of resettlement and rehabilitation processes in seven tiger reserves after 2007 – i.e. following the promulgation of the FRA and the amendments to the Wildlife Protection Act and the resettlement compensation package – found that at several sites people had not been told about the option of staying in the respective tiger reserve (Shahabuddin and Bhamidipati, 2014). Administrators had ignored provisions in the Forest Rights Act; for instance, they declared that the communities caused irreversible damage to the wildlife without providing any evidence. Only in two out of seven tiger reserves had the consent of individual families to relocate been recorded – despite the resettlement being acknowledged as voluntary. At the five other sites, there was at least some form of coercion or pressure from forest officials to relocate (Shahabuddin and Bhamidipati, 2014).

One of the most extensively studied resettlement cases is the Similipal Tiger Reserve (STR) in the state of Odisha (formerly Orissa). Similipal was declared as a 'Tiger Reserve' under the late Prime Minister Indira Gandhi's national flagship conservation programme 'Project Tiger' in the year 1973. Six years later the Orissa state government declared it a wildlife sanctuary with a designated area of 2,750km^2. In 2009, STR was included into the World Network of Biosphere Reserves by UNESCO. The 'core zone' of the STR (1,194km^2) is a designated national park although the central government has yet to issue a final notification due to the fact that three villages have not been relocated to date (Dash and Bahera, 2018). There are 61 villages in the 'buffer zone' of the reserve where some forms of ecotourism are allowed.

The majority of the forest-dwelling population in the STR belong to the category of 'scheduled tribes', a term that denotes Indigenous peoples with a formally acknowledged status as 'historically disadvantaged groups.' Some of these groups are categorised as 'primitive' by the state government of Odisha (Dash and Bahera, 2018). Yet the communities insist that their livelihood strategies are not a threat to the forest and its wildlife and that they can co-exist peacefully with nature.

> We are doing no harm to the forest and wild animals. Rather we worship them. We are living in their home. So, we should love and respect them. However, the Government has always treated us as the enemies of nature.
> (Village head in the STR core zone, cited by
> Dash and Bahera, 2018, p. 332)

The three villages that remain in the core zone face a range of restrictions, including a prohibition on the collection of valuable non-timber forest products (NTFPs) – their major livelihood source. While one community has been categorically opposed to relocation, another was ready to leave the tiger reserve after forest officials showed them a proposed settlement area suitable for agriculture. Despite this, villages were later told that the proposed area had been earmarked for mining and were presented with an alternative site unsuitable for farming. This led to their rejection of the proposed relocation.

The last relocation of an entire indigenous community from the STR core zone occurred in 2010, in disregard of settlement claims that villagers had filed earlier. Villagers were presented with two options, as per the Forest Rights Act: option 1 involved a one-off cash payment (without land provision and any further livelihood support), whereas option 2 required the Forest Department to relocate the villagers from the tiger reserve's core zone to a 'rehabilitation colony' located 50km away from their original residential area (Shahabuddin and Bhamidipati, 2014). For the two-thirds of villagers who chose option 2 the relocation meant better access to infrastructure, such as education, electricity and health facilities. Yet, they suffered from high livestock mortality, lower agricultural productivity, reduced access to fodder, NTFPs and firewood, extreme heat conditions during summer, and a higher dependency on wage labour. Nearly all resettled villagers under option 2 reported negative impacts on their cultural values and community life (Dash and Bahera, 2018). The fate of the villagers who chose option 1 remains unknown.

Conservation through militarisation, tourism and eviction in Tayrona National Park, Colombia

The Tayrona National Park (*Parque Nacional Natural Tayrona*) is located on the Caribbean coast of northern Colombia and is one of 60 protected areas under the South American country's National Park System that covers around 17 million hectares or around 15 per cent of the national territory. The park was declared a protected area as early as 1964 and covers an area of 15,000 hectares, including a 3,000-hectare marine area (Brüggemann and Rodríguez, 2004). It is one of the most biodiverse areas in Colombia. Popular with both domestic and foreign tourists, the park welcomed nearly 450,000 visitors in 2018 (Álvarez, 2019).

While the Tayrona National Park does not overlap with any designated Indigenous reservations (*resguardos*), it is part of the ancestral territory of four Indigenous groups that inhabit the nearby Sierra Nevada de Santa Marta, which was declared a Biosphere Reserve in combination with the Tayrona National Park by UNESCO in 1982 (Ojeda, 2012). The 30,000 members of these Indigenous groups, the Kogui, Wiwa, Arhuaco and Kankuamo, who live in the combined Biosphere Reserve area of 400,000 hectares, trace their ancestry back to the pre-Colombian Indigenous Tayrona from whom the park derives its name. Sacred places – where the Indigenous groups present annual *pagamentos* (ritual offerings) to *pacha mama* (Mother Earth) – are recognised in the park's management plan (Bocarejo and Ojeda, 2016). The local office of the *Unidad Administrativa Especial de Parques Nacionales Naturales* (UAESPNN) in charge of park management has aimed at forging close ties with Indigenous associations who they regard as eco-guardians and conservation allies (Ojeda, 2012; Bocarejo and Ojeda, 2016).

The Tayrona National Park is inhabited by several small fishing communities and many peasant groups that have migrated to the park since the middle of the 20th century (Ojeda, 2012). Most of these settlers (*colonos*) were displaced by violent conflicts in other parts of the country, and many of them have been involved in the cultivation of illicit crops, such as coca and marihuana (Ojeda, 2012). In contrast to the Indigenous groups that inhabit primarily the Sierra Nevada de Santa Marta protected area, the *colonos* are stigmatised as both eco-threats and security threats. From the 1970s until the early 2000s, the park's area was controlled by paramilitary forces – with knowledge of and support from the Colombian state – that provided security to drug lords and private landowners who had become the targets of guerrilla groups (Bocarejo and Ojeda, 2016). Land ownership of the Tayrona National Park has been opaque and complex; while Ojeda (2012, p. 363) maintains that 90 per cent of the protected area is "*de facto* in private hands … notwithstanding its legal public character", media reports suggest that 84 per cent of the park are 'properties under discussion', hence the exact legal status of the majority of the park remains unknown (*El Espectador*, 2017). Some of the large landowners are local elites whose ancestors obtained the properties through royal decrees dating back to Spanish colonial times (i.e. pre-1819) or who have recently acquired land titles despite the illegality of land sales given the protected area status of the park (Ojeda, 2012).

Shortly after taking office, former president Álvaro Uribe (2002–2010) launched a national policy of 'Democratic Security'. Part of that policy was a dual strategy of tourism promotion and securitisation of tourism hotspots to guarantee the mobility and relative safety of tourists (Ojeda, 2012). The strategy culminated in the promotional campaign 'Colombia is Passion' with the slogan 'Colombia: The only risk is wanting to stay' kicked off in late 2007 by the government's tourism promotion arm, Proexport Colombia (Fletcher, 2011). For the Tayrona National Park and its inhabitants, this meant being increasingly targeted by paramilitary 'clean-up' operations. Given the insecurity of land tenure within the park, the peasant settlers (*colones*) and fishers could be easily dispossessed and evicted from their land to 'secure the area' and make room for tourists. In March 2010, a seven-household strong fishing community was evicted by park officials who destroyed the fishers' huts that had existed for about 50 years, while sparing the luxurious private beach homes (Bocarejo and Ojeda, 2016). Most fishers had to move to the nearby city of Santa Marta to look for new livelihood opportunities (Ojeda, 2012).

A more subtle approach was employed in 2005 through granting a ten-year tourism concession to Unión Temporal Tayrona (UTT), an alliance between the Chamber of Commerce of the nearby city Santa Marta, the private travel company Aviatur and the travel agency Alnuva, with the aim of relieving the UAESPNN of the burden of administering tourism activities in the park (Ojeda, 2012). Although UTT occupies not more than one per cent of the park's territory, it controls the strategically important tourist zones along with the two park entrances. Ojeda's (2012) study found that the privatisation of strategic areas through UTT has increased the pressure on natural resources, park inhabitants and local associations who formerly provided most of the tourist services.

Uribe's successor in the Colombian presidency, Juan Manuel Santos (2010–2018) – who would later become a recipient of the Nobel Peace Prize 2016 for his efforts in negotiating a peace agreement with Colombia's largest rebel group, the Revolutionary Armed Forces of Colombia (*Fuerzas Armadas Revolucionarias de Colombia* – FARC) – further stepped up the promotion of tourism in Tayrona National Park by attempting to tap into the high-end tourist market. In 2011, plans were revealed that the Thai multinational Six Senses Hotels, Resorts & Spas group wanted to construct a luxury resort on one of the park's pristine beaches (Villegas, 2019). The project was backed by President Santos who praised the company's ecological virtues, yet it was alleged in the Colombian media that some of his family members had been involved in early design phases (Bocarejo and Ojeda, 2016). As part of the legal process, the Ministry of Interior had to declare whether Indigenous groups were present in the area (*El Heraldo*, 2011; Bocarejo and Ojeda, 2016) which would necessitate a prior consultation process according to Colombia's multicultural legislation and the country's commitment to the United Nations Declaration for the Rights of Indigenous Peoples (UNDRIP). When the Ministry certified that there was no indigenous presence in the park boundaries, Indigenous communities were outraged as they have evidence of historical ties to the land through earlier settlement, occasional transit and regular rituals (Bocarejo and Ojeda, 2016). While in 2012 it was reported that three of the four Indigenous groups present in the area had given their approval to the tourism project, the Kankuamo remained categorically opposed, stating that it infringed on their sacred sites and that their position was "untouchable and non-negotiable" (Trent, 2012, n.p.). Following continuing pressure from Indigenous associations, local and national politicians, legal representatives, and environmental groups, President Santos himself terminated the project but with the stern warning that he would "cleanse the park" (Bocarejo and Ojeda, 2016, p. 180). The Six Senses project was not the only luxury ecotourism project, where Santos had to change his original stance. A project led by a domestic investor linked with the travel company Aviatur, the Los Ciruelos Eco-Resort project, also had to be stopped in 2013 with massive implications including a multi-million US$ lawsuit (see Box 7.4).

Box 7.4 Attempted green grab by the ecotourism project of the Colombian company *Reserva Los Ciruelos* in Tayrona National Park

In 2009, the Colombian company Reserva Los Ciruelos with close ties to the travel company Aviatur was granted a government permit to construct a luxury resort with twelve eco-lodge units in Tayrona's Concha Bay. Yet it was later revealed that the construction would affect one of the park's important ecosystems, a biodiversity-rich dryland forest, and infringe on the sacred sites of Indigenous people who were not previously consulted about the project.

Construction of the eco-resort started in 2011 but was suspended later that year, when the National Environmental Licensing Authority (*Autoridad Nacional de Licencias Ambientales* – ANLA) and UAESPNN found out that the company had withdrawn freshwater from an illegal source. The suspension was lifted in late 2012, but in the meantime two research institutions (the Humboldt Institute in Bogota and Icesi University in Cali) had informed the environmental authorities that the dry forest where the eco-resort was to be located would suffer irreparable damage. In early 2013, then President Santos expressed his disapproval of the Los Ciruelos project via Twitter. It was later revealed that he had been warned as early as November 2011 that the construction of hotels would constitute a threat to Tayrona's ecosystems. Following another temporary stop to the project imposed by ANLA in mid-January 2013 on environmental grounds, the Magdalena Administrative Court ordered the immediate suspension of Los Ciruelos environmental license until proper consultations with the four Indigenous communities be conducted, a ruling that was upheld a year later by the State Council.

The company filed a lawsuit against the State in 2014 for direct reparations of US$3.37 million, followed by another one in 2015 against ANLA, the Humboldt Institute and UEASPNN for compensation of US $180,000 in lost revenues. In mid-2015, ANLA lifted the suspension and issued a modified environmental license, but – according to the company – the regulations had changed the economic parameters of the project to such an extent that it was not considered financially viable anymore.

Sources: Edmond, 2013; Schertow, 2013; *El Espectador*, 2013, 2017; O'Gorman, 2017; Álvarez, 2019; Segiumento, 2019

The judgement against the Los Ciruelos Eco-Resort project also had major implications for Aviatur's other tourism services, as the travel company was ordered to consult with Indigenous communities regarding all activities that affect their sacred sites. In the years 2017–2019, the Tayrona National Park was closed for one month annually to give the local ecosystem time to 'heal' from the impact of tourism. In mid-2019, a public tender was issued, as UTT's concession would expire in 2020 after a 15-year lease (Álvarez, 2019). Yet, in December of the same year, UAESPNN suspended the bidding process, awaiting a decision by the Administrative Court of Santa Marta on the guarantee of the fundamental right to prior consultation with the Indigenous communities (Flórez Arias, 2019). As the next tourism concession will have a duration of 23 years, the Tayrona National Park and its guardians have reached a critical juncture.

Concluding remarks

This chapter has examined the nexus of wildlife tourism, fortress conservation and forced displacement. Opening protected areas to domestic visitors and foreign tourists provides legitimacy for demarcating and securitising these areas and removing any former residents that are considered 'threats' to an imaginary 'wilderness'. All four case studies provide evidence that the protection and commodification of nature for tourist consumption can have detrimental impacts on local communities that have co-existed with wildlife long before these protected area enclaves were opened for tourism. While in Tanzania the government has run an openly hostile campaign against Indigenous peoples for many years and evicted hundreds of pastoralist communities from wildlife reserves such as the Serengeti National Park (often backed by international wildlife conservation organisations), the approach taken by the Mozambiquan authorities to resettle communities from the newly established Limpopo National Park was more subtle, dubbed as 'induced volition' (Milgroom and Spierenburg, 2008). Similar strategies have been adopted by park authorities in India after the introduction of new legal frameworks related to wildlife protection and resettlement, particularly the promulgation of the Forest Rights Act in 2006. In Colombia's Tayrona National Park, a triple strategy of privatisation, ecotourism promotion and militarisation of tourist spots and travel routes has been employed by local elites and paramilitary forces to 'protect pristine nature' from the destructive action of 'invaders' and 'illegal occupants' who have been largely deprived of their customary access to the Park's resources (cf. Ojeda, 2012).

The next chapter explores processes of gradual displacement and gentrification associated with (eco)cultural heritage tourism in the rural and urban contexts of Central America (Guatemala), East Asia (China), Southeast Asia (Cambodia) and South America (Argentina and Peru). These case studies will show the convergence of political agendas surrounding tourism development and historical heritage preservation and how this process undermines local people's property and housing rights as well as their rights to participate meaningfully in the management of (eco)cultural heritage sites.

References

Adams, J.S. and McShane, T.O. (1992) *The Myth of Wild Africa: Conservation without Illusion.* W.W. Norton: New York.

Akama, J.S. (2004) 'Neocolonialism, dependency and external control of Africa's tourism industry: A case study of wildlife safari tourism in Kenya' in Hall, C.M. and Tucker, H. (eds) *Tourism and Postcolonialism: Contested Discourses, Identities and Representations.* Routledge: London, New York, pp. 140–152.

Benjaminsen, T.A. and Bryceson, I. (2012) 'Conservation, green/blue grabbing and accumulation by dispossession in Tanzania', *Journal of Peasant Studies*, 39 (2): 335–355.

Benjaminsen, T.A., Goldman, M.F., Minwary, M.Y. and Maganga, F.P. (2013) 'Wildlife management in Tanzania: State control, rent seeking and community resistance', *Development and Change*, 44 (5): 1087–1109.

Bluwstein, J., Lund, J.F., Askew, K., Stein, H., Noe, C., Odgaard, R., Maganga, F. and Engström, L. (2018) 'Between dependence and deprivation: The interlocking nature of land alienation in Tanzania', *Journal of Agrarian Change*, 18: 806–830.

Bocarejo, D. and Ojeda, D. (2016) 'Violence and conservation: Beyond unintended consequences and unfortunate coincidences', *Geoforum* 69: 176–183.

Brockington, D. and Igoe, J. (2006) 'Eviction for conservation: A global overview', *Conservation & Society*, 4 (3): 424–470.

Brockington, D., Duffy, R. and Igoe, J. (2008) *Nature Unbound: Conservation, Capitalism and the Future of Protected Areas.* Earthscan: London.

Brüggemann, J. and Rodríguez, E.E. (2004) 'Tayrona National Park, Colombia: International support for conflict resolution through tourism', *Parks*, 14 (1): 40–47.

Burgoyne, C. and Mearns, K. (2017) 'Managing stakeholder relations, natural resources and tourism: A case study from Ololosokwan, Tanzania', *Tourism and Hospitality Research*, 17 (1): 68–78.

Busscher, N., Parra, C. and Vanclay, F. (2018) 'Land grabbing within a protected area: The experience of local communities with conservation and forestry activities in Los Esteros del Iberá, Argentina', *Land Use Policy*, 78: 572–582.

Carruthers, J. (1995) *The Kruger National Park. A Social and Political History.* University of Natal Press: Pietermaritzburg.

Dash, M. and Bahera, B. (2018) 'Biodiversity conservation, relocation and socio-economic consequences: a case study of Similipal Tiger Reserve, India', *Land Use Policy*, 78: 327–337.

DeMotts, R. (2017) *The Challenges of Transfrontier Conservation in Southern Africa: The Park Came After Us.* Lexington Books: Maryland.

Forsyth, T. and Walker, A. (2008) *Forest Guardians, Forest Destroyers: The Politics of Environmental Knowledge in Northern Thailand.* Silkworm Books: Chiang Mai.

Gardner, B. (2012) 'Tourism and the politics of the global land grab in Tanzania: Markets, appropriation and recognition', *Journal of Peasant Studies*, 39 (2): 377–402.

Hill, A. (2017) 'Blue grabbing: reviewing marine conservation in Redang Island Marine Park, Malaysia', *Geoforum* 79: 97–100.

Hussain, A., Dasgupta, S. and Singh Bargali, H. (2016) 'Conservation perceptions and attitudes of semi-nomadic pastoralist towards relocation and biodiversity management: A case study of Van Gujjars residing in and around Corbett Tiger Reserve, India', *Environment, Development and Sustainability*, 18 (1): 57–72.

IWGIA (2019a) *Annual Report 2019.* International Work Group for Indigenous Affairs (IWGIA): Copenhagen.

IWGIA (2019b) *The Indigenous World 2019*. International Work Group for Indigenous Affairs (IWGIA): Copenhagen.

Lascorgeix, A. and Kothari, A. (2009) 'Displacement and relocation of protected areas: A synthesis and analysis of case studies', *Economic & Political Weekly*, 44: 37–47.

Lindsey, P.A., Balme, G.A., Booth, V.R. and Midlane, N. (2012) 'The significance of African lions for the financial viability of trophy hunting and the maintenance of wild land', *PLoS ONE*, 7 (1): e29332. doi:10.1371/journal.pone.0029332.

Lunstrum, E. and Ybarra, M. (2018) 'Deploying difference: security threat narratives and state displacement from protected areas', *Conservation and Society*, 16 (2): 114–124.

Mahapatra, A.K., Tewari, D.D. and Baboo, B. (2015) 'Displacement, deprivation and development: The impact of relocation on income and livelihood of tribes in Similipal Tiger and Biosphere Reserve, India', *Environmental Management*, 56: 420–432.

Massé, F. and Lunstrum, E. (2016) 'Accumulation by securitization: Commercial poaching, neoliberal conservation, and the creation of the new wildlife frontiers', *Geoforum*, 69: 227–237.

McClanahan, T.R. and Mangi, S. (2000) 'Spillover of exploitable fishes from a marine park and its effect on the adjacent fishery', *Ecological Applications*, 10: 1792–1805.

Milgroom, J. and Spierenburg, M. (2008) 'Induced volition: Resettlement from the Limpopo National Park, Mozambique', *Journal of Contemporary African Studies*, 28: 435–448.

Mowforth, M. and Munt, I. (2016) *Tourism and Sustainability: Development, Globalisation and New Tourism in the Third World* (4th edition). Routledge: London, New York.

Nelson, F. and Blomley, T. (2010) 'Peasants' forests and the King's game? Institutional divergence and convergence in Tanzania's forestry and wildlife sector', in Nelson, F. (ed.) *Community Rights, Conservation & Contested Land*. Earthscan: London, Washington DC, pp. 79–105.

Ngoitiko, M., Sinandei, M., Meitaya P. and Nelson, F. (2010) 'Pastoral activists: Negotiating power imbalances in the Tanzanian Serengeti', in Nelson, F. (eds) *Community Rights, Conservation & Contested Land*. Earthscan: London, Washington, DC, pp. 269–289.

Ojeda, D. (2012) 'Green pretexts: Ecotourism, neoliberal conservation and land grabbing in Tayrona National Natural Park, Colombia', *The Journal of Peasant Studies*, 39 (2): 357–375.

Otsuki, K., Achá, D. and Wijnhoud, J.D. (2017) 'After the consent: Re-imaging participatory land governance in Massingir, Mozambique', *Geoforum*, 83: 153–163.

Paterniti, M. (2017) 'Should we kill animals to save them?', *National Geographic*, 232 (4): 70–99.

Pearce, F. (2012) *The Land Grabbers: The New Fight over who Owns the Earth*. Bacon Press: Boston, MA.

Porokwa, E. (2018) 'Tanzania', in IWGIA (ed.) *The Indigenous World in 2018*. International Work Group for Indigenous Affairs (IWGIA): Copenhagen, pp. 482–492.

Quammen, D. (2016) 'The paradox of the park', *National Geographic*, 229 (5): 54–91.

Rodgers, G. (2009) 'The faint footprint of man: representing race, place and conservation on the Mozambique–South Africa Borderland', *Journal of Refugee Studies*, 22 (3): 392–412.

Shahabuddin, G. and Bhamidipati, P.L. (2014) 'Conservation-induced displacement: Recent perspectives from Indi', *Environmental Justice*, 7 (5): 122–129.

Witter, R. and Satterfield, T. (2014) 'Invisible losses and the logics of resettlement compensation', *Conservation Biology*, 28 (5): 1394–1402.

Media sources/websites

Álvarez, N.N. (2019) 'Las batallas del parque Tayrona, más allá de la concesión [The battles of Tayrona Park, beyond the concession]', 30 August. Available at: www.eltiempo.com/vida/viajar/como-sera-la-nueva-concesion-en-el-parque-tayron a-406700 (accessed 23 June 2020).

Ayyar, K. (2018) 'How India's conservationists are fighting to save half of the world's tigers', *Time Magazine*, 28 July. Available at: http://time.com/5345610/globa l-tiger-day-tigers-india-conservation/ (accessed 18 December 2018).

Edmond, R. (2013) 'Colombia's environment minister expresses concerns over tourism plans in national park', *Colombia Reports*, 11 January. Available at: https://colombia reports.com/amp/colombias-environment-minister-expresses-concerns-over-tourism -plans-in-national-park/ (accessed 23 June 2020).

El Espectador (2013) 'Fallo prohíbe el turismo en sitios sagrados del Tayrona [Judgement prohibits tourism in Tayrona's sacred sites]', 14 February. Available at: www.elespec tador.com/noticias/nacional/fallo-prohibe-el-turismo-en-sitios-sagrados-del-tayrona/ (accessed 23 June 2020).

El Espectador (2017) 'Los Ciruelos pide más de $6.000 millones por no haber podido construir en el Tayrona [Los Ciruelos asks for more than $ 6,000 million for not having been able to build in Tayrona]', 28 July. Available at: www.elespectador. com/noticias/judicial/los-ciruelos-pide-mas-de-6000-millones-por-no-haber-podido -construir-en-el-tayrona/ (accessed 23 June 2020).

El Heraldo (2011) 'El hotel Six Senses Tayrona: Un anuncio con muchos peros [The Six Senses Hotel Tayrona: An announcement with many buts]', 17 October. Available at: www.elheraldo.co/region/el-hotel-six-senses-tayrona-un-anuncio-con-muchos-p eros-42151 (accessed 23 June 2020).

Flórez Arias, J.M. (2019) 'Suspenden concesión turística del Tayrona [Tayrona tourism concession suspended]', *El Colombiano*, 19 December. Available at: www.elcolombia no.com/colombia/licitacion-turistica-del-tayrona-es-suspendida-por-parques-naturales-A F12172345 (accessed 24 June 2020).

O'Gorman, J. (2013) 'Tayrona eco-tourism stopped again', *Colombia Reports*, 13 February. Available at: https://colombiareports.com/tayrona-eco-tourism-stopped-aga in/ (accessed 23 June 2020).

Schertow, J.A. (2013) 'Colombia: Court suspends eco-tourism project in Tayrona National Park', 18 February. Available at: https://intercontinentalcry.org/colombia -court-suspends-eco-tourism-project-in-tayrona-national-park/ (accessed 23 June 2020).

Seguimiento (2019) 'Sigue el pleito legal de Los Ciruelos sobre construcción en el Tayrona [The legal lawsuit of Los Ciruelos on construction in Tayrona continues]', 4 February. Available at: https://seguimiento.co/la-samaria/sigue-el-plei to-legal-de-los-ciruelos-sobre-construccion-en-el-tayrona-22161 (accessed 24 June 2020).

Survival International (n.d.) 'Background briefing – Parks and peoples'. Available at: www.survivalinternational.org/about/parks-and-peoples (accessed 31 December 2018).

Trent, C. (2012) 'Indigenous divided on construction of Tayrona resort', *Colombia Reports*, 31 October. Available at: https://colombiareports.com/indigenous-divide d-on-construction-of-tayrona-resort/ (accessed 23 June 2020).

Vidal, J. (2016) 'The tribes paying the brutal price of conservation', *The Guardian*, 28 August. Available at: www.theguardian.com/global-development/2016/aug/28/exiles-human-cost-of-conservation-indigenous-peoples-eco-tourism (accessed 31 December 2018).

Villegas, I..V. (2019) 'Parque Nacional Natural Tayrona', 1 March. Available at: https://prezi.com/p/tj-8ohqgba0m/conflicto-ambiental-parque-nacional-tayrona/ (accessed 23 June 2020).

8 Cultural heritage tourism

Beautification, gentrification, eviction

There is no universally accepted definition of 'heritage tourism', but most authors refer to such notions as (eco-)cultural heritage with both tangible and intangible elements that attract domestic and international visitors (Ashworth, Graham and Tunbridge, 2007; Timothy, 2015). While humankind's cultural heritage is often "mundane, commonplace and very personal to individuals and local communities", most tourists "seek cultural sites that are world-renowned, tangible and very old" (Timothy, 2015, p. 238). Policies surrounding such 'global' heritage sites are often entangled with relations of power, exercised in subtle forms in some cases, while using brute state force in others (Silverman and Ruggles, 2007).

Mowforth and Munt (2016, p. 376) quote the Swedish International Development Agency's suggestion that 'heritage tourism' – when done in a sensible and sensitive manner – can be considered the urban equivalent of 'ecotourism'. Yet the difference between cultural and natural heritage sites is not clear-cut and remains highly contested. The United Nations Educational, Scientific and Cultural Organization (UNESCO) listed more than 1,120 properties as World Heritage Sites as of July 2020, out of which the organisation classified 869 as cultural heritage, 213 as natural heritage and 39 as mixed cultural/natural heritage (https://whc.unesco.org/en/list/). The International Work Group for Indigenous Affairs (IWGIA) and the Forest Peoples Programme (FPP) have criticised the differentiation between cultural and natural heritage as "artificial and problematic in the case of World Heritage Sites located in indigenous people's territories, because the lives, cultures and spiritual beliefs of indigenous peoples are inseparable from their lands, territories, and natural resources" (IWGIA and FPP, 2015, pp. 5–6). Indigenous peoples have also expressed their concerns over the "lack of regulations to ensure meaningful participation and free, prior and informed consent of indigenous peoples in the nomination and designation of World Heritage sites" (ibid., pp. 10–12). Similar concerns have been raised about the nearly 700 biosphere reserves located in 122 countries, which are designated by the UNESCO as "learning sites for sustainable development" (UNESCO, n.d.).

Apart from the lack of local participation in heritage development, another problem that is typical for World Heritage Sites is the imposition of an international legal regime on national and local legal frameworks. World Heritage designations by the UNESCO come with the prerequisites that the state bears the responsibility for (1) the enactment of protective legislation, (2) the establishment of permanent boundaries around the site and (3) the definition of meaningful buffer zones, which can result in overruling the customary or formalised rights of residents in World Heritage Sites (Gillespie, 2009). In urban heritage sites, such as historical centres in Latin American's major cities, gentrification, eviction and cultural dispossession have become common phenomena (Janoschka and Sequera, 2016).

The following sections will examine the controversies surrounding cultural heritage preservation, eco-cultural tourism and securitisation in Guatemala, the effects of cultural heritage tourism on Indigenous and non-indigenous people's rights and livelihoods in China, land grabbing and resettlement in the Angkor Archaeological Park in Cambodia, and gentrification and eviction processes in urban heritage sites in Argentina and Peru.

Cultural heritage preservation, border securitisation and eco-cultural tourism in Guatemala's Maya Biosphere Reserve

The Maya Biosphere Reserve (MBR) is located in Petén in northern Guatemala and was identified by the state-run Guatemalan Tourism Commission as 'The Heart of the Mayan World' for two major tourism campaigns that started in the early 2000s. The MBR was established in 1990 by Guatemalan and foreign conservationists in the midst of the country's decade-long peace process (Devine et al., 2020). Covering an area larger than El Salvador (Guatemala's neighbour to the south), it is the largest protected area in Central America and provides livelihoods for an estimated 180,000 people (Cuffe, 2016). The MBR is a complex patchwork of national parks – most of which are located in the western part of the reserve – and a large multiple use zone that dominates its eastern part, divided into several community-based concessions (Rahder, 2015). Most residents were not even aware of the creation of the MBR until after the laws had been enacted (Lunstrum and Ybarra, 2018). Many of them experienced the declaration of the MBR "as an act of enclosure and land dispossession" (Devine et al., 2020, p. 1023). The MBR is home to several hundred archaeologically important sites, including the UNESCO World Heritage Maya site of Tikal, which receives hundreds of thousands of domestic and foreign visitors each year (Devine, 2017). While UNESCO has recorded negative impacts of tourism on Tikal, such as waste, pollution and vandalism, many other sites remain virtually unexplored due to accessibility and security issues.

Over the past decade, the government of Guatemala has promoted tourism in these largely unvisited areas, through such initiatives as the Four Jaguar Eco-Tourism Project which ran from 2008 to 2012 (Devine, 2014). However, the promotion of cultural and eco-tourism for economic growth appears to be only part of the government's agenda. Another major goal is to securitise the area which has allegedly been a site of illicit timber extraction, poaching, drug trafficking and antiquity looting and whose porous borders with neighbouring Belize and Mexico still remain contested (Devine, 2017). The military plays a major role in this effort, and several military outposts have been established since 2008 in the MBR, which has been the focal area of new tourism development, oftentimes by evicting village communities that are branded as insurgents and eco-threats (Devine, 2014). Meanwhile, several village cooperatives inside the biosphere's multiple use zone depend on collectively managed forest concessions and community-based tourism for their livelihoods (Devine et al., 2020).

In late 2019, the Guatemalan government extended the 25-year license for the first and one of the largest community concessions, the Carmelita community (Pearce, 2020). Yet Indigenous communities continue to be framed as "forest-eating 'termites'" by non-indigenous elites (Lunstrum and Ybarra, 2018, p. 119). Countering this narrative, local communities and community tourism organisations argue that if the forest concessions did not exist, the whole MBR would have already been destroyed. This view has been supported by the Guatemala's National Council for Protected Areas (CONAP), the United States Agency for International Development (USAID), and several environmental NGOs that maintain that the communally managed forest concessions provide an effective buffer against the expansion of farming and cattle ranching (Cuffe, 2016; Pearce, 2020).

Forest concessions have benefitted from close collaboration with international NGOs, such as Rainforest Alliance, that help them in finding markets for non-timber forest products (NTFPs), such as chicle, a latex from the sapodilla tree, that can be sold in international fair trade and organic markets (Pearce, 2020). A recent study also contented that communities involved in forest concessions have a genuine interest in maintaining intact forests to sustain their income from sustainably harvested wood, NTFPs and community-based tourism (Bocci et al., 2018). This view has been confirmed in a study by Devine et al. (2020) who found that in the eastern parts of the MBR – which are primarily occupied by villages and community forestry concessions – forests have remained largely intact, while national parks in the western part of the reserve have experienced some of the highest deforestation rates in Latin America and globally, despite much stricter regulations which only allow conservation and tourism activities. The study's authors attribute forest destruction in the MBR primarily to the phenomenon of 'narco-cattle ranching' which they define as illicit activities that combine large-scale cattle farming with drug trafficking and money laundering (Devine et al., 2020). It is alleged that the narco-ranchers in MBR are funded by two Mexican drug cartels that are also involved in forest conversion to cattle ranches across the border in Mexico's Calakmul Biosphere Reserve (Pearce, 2020).

Different stakeholder groups (archaeologists, local communities, private corporations, government officials and environmental NGOs) have diverging opinions on how to move forward with tourism, development and conservation in the MBR (Cuffe, 2016). A US American archaeologist, Dr Richard Hansen, has proposed to replace logging-based livelihoods with cultural tourism and pharmaceutical industry bioprospecting (Cuffe, 2016). Dr Hansen's Foundation for Anthropological Research and Environmental Studies (FARES) is supported by Guatemala's state archaeological institute, IDAEH (*Institut de Antropología e Historia*), the United States Department of the Interior (DOI), Idaho State University, a former Guatemalan President and US American actor Mel Gibson (Rahder, 2015). His Mirador Basin Project has received funding from National Geographic and is backed by Pacunam, a Guatemalan corporate foundation which includes a powerful local cement company, multinational corporation Walmart and several banks and agro-industry companies, and by the US-based NGO 'Global Heritage Fund' (Cuffe, 2016).

Community tourism representatives and the association representing 22 community forestry organisations in Peten (*Asociación de Comunidades Forestales de Petén* – ACOFOP) strongly oppose Dr Hansen's and Pacunam Foundation's plan to connect Mirador with other Mayan sites in the MBR through improved road infrastructure to facilitate mass tourism, as they believe this would jeopardise community-based tourism activities and threaten the integrity of existing forest concessions (Rahder, 2015; Cuffe, 2016). Yet, in 2019, Dr Hansen forged a powerful alliance with a group of influential US senators who proposed to US Congress a bill (S. 3131) to authorise "the Secretary of the Interior to establish a Maya Security and Conservation Partnership program" (US Congress, 2019, p. 1). Under the proposed partnership the Guatemalan government would receive US$60 million "to fund field-based tropical and archaeological research, law enforcement, and sustainable tourism activities within the [Mirador] basin" (US Congress, 2019, p. 3), with matching funds in the form of a loan of US$60 million sought from the Central American Bank for Economic Integration (*Banco Centroamericano de Integración Económica* – BCIE) (Maya Biosphere Watch, 2020).

Critics of the Maya Security and Conservation Partnership program fear that it will turn Mayan cultural heritage in Guatemala into a US-funded theme park for American tourists (García, 2020). While local communities have yet to be formally consulted about the project, community-based website Maya Biosphere Watch and critical media outlets allege that the 'Partnership' does not reveal its real private-sector driven agenda and that the 'sustainable tourism model' will be detrimental to the fragile ecosystem of the MBR (Maya Biosphere Watch, 2020; García, 2020). Indeed, the plan is to build large hotels and a train network to allow for increased and potentially unsustainable tourist activities. There are also allegations that portions of the forest concessions would be absorbed into national parks if the plans of the Maya Security and Conservation Partnership program would go ahead (Pearce, 2020).

Eco-cultural heritage tourism, involuntary resettlement and livelihood restrictions in China

As of August 2020, Mainland China hosted 37 properties listed as world cultural heritage sites, 14 as natural heritage sites and four as mixed cultural/

natural heritage sites by the UNESCO, most of which attract a large number of domestic and foreign visitors (https://whc.unesco.org/en/list/). As Gao, Lin and Zhang (2020) contend, cultural heritage rights and responsibilities in China are complicated by the fact that local communities that are supposed to be caretakers of heritage sites do not have long-term tenure security and can be expropriated and relocated by the state at any time for the 'public good'. Hence, World Heritage Site listings in Mainland China are often associated with the relocation of populations and/or restrictions on certain livelihood activities. Many of them have also been the source of intense conflict between government, corporate interests and local communities (Zhang, Fyall and Zheng, 2015; Li, Lau and Su, 2020).

Mount Sanqingshan National Park in Jiangxi Province – designated as a World Heritage Site in July 2008 primarily due its uniquely shaped granite pillars and peaks – is one example of a 'natural heritage' that triggered the forced resettlement of communities. Over the first five years following its designation, tourist numbers more than quadrupled and about 1,300 residents living in close proximity to the national park were relocated and most of their agricultural fields were acquired by the state for tourism site development (Su, Wall and Xu, 2016). While traditional (pre-resettlement) livelihood activities had been diverse and composed of crop cultivation, animal husbandry, forestry and migration prior to the development of heritage tourism, tourism-related activities became by far the most dominant livelihood source in the village after the relocation (Su, Wall and Xu, 2016). Tourism has been widely considered as bringing short-term economic benefits, yet it also rendered resettled communities more dependent on a single source of income, which may jeopardise their long-term livelihood sustainability (Su, Wall and Xu, 2016).

A World Heritage Site that was listed by UNESCO at the same time as the Mount Sanqingshan National Park is the Hongkeng *tulou* cluster in Fujian Province. The *tulou* (earth buildings) are large multi-story buildings with tall fortified mud walls that were originally built for defensive purposes and can provide home for a few hundred people (Wang and Yotsumoto, 2019). In order to obtain world heritage status, the county government demolished modern houses, expropriated land and prohibited the building of new houses without providing suitable alternatives to cope with high population growth. Over a period of ten years, 3.7 hectares of physical constructions were demolished that were not considered in harmony with the *tulou* buildings and close to 1,500 local residents were relocated, according to a study by Li, Lau and Su (2020). Many villagers lost up to 70 per cent of their farmland in the process with very little monetary compensation (Wang and Yotsumoto, 2019).

The building ban imposed severe restrictions on young people who mostly prefer modern houses over traditional ones. An interviewee from a local government body in Li, Lau and Su's (2020) study expressed frustration about residents building new houses inside the protected area and converted heritage houses into shops. Following the World Heritage designation, street vending – the major alternative to farming was prohibited in the name of 'beautification' for tourism, and the ban was enforced by damaging or confiscating vendors' stands (Wang and Yotsumoto, 2019). Meanwhile, some *tulou* owners expressed their satisfaction with increased tourism, as they annually receive a fixed rent for their properties from the government and eight per cent of the revenues from the entrance tickets (Li, Lau and Su, 2020).

One of the latest additions to China's long list of World Heritage Sites is the Cultural Landscape of Honghe Hani Rice Terraces in Southern Yunnan Province, designated by UNESCO in 2013. The world heritage status has created additional momentum to an already increasing tourism stream to the area. The three most visited Indigenous Hani villages are located in the protected zone of the WHS and are subject to a number of conservation regulations, such as prohibitions to abandon the terraces, maintenance of a 'traditional' housing style that adheres to principles of 'heritage authenticity and integrity', as well as rules pertaining to forest and water management (Gao, Lin and Zhang, 2020). Adherence to these regulations is complicated by a high degree of out-migration of younger Hani in search for better opportunities elsewhere and the fact that some Indigenous villagers have agreed to renting out their homes to external (majority Han Chinese) entrepreneurs who turned them into guest-houses and restaurants (Chan et al., 2016). The newcomers do not have an emotional connection to the cultural landscape and feel little responsibility for contributing to its preservation (Gao, Lin and Zhang, 2020). Meanwhile, villagers who ignored the government regulations for 'authentic' construction faced demolition of their houses by armed police (Gao, Lin and Zhang, 2020). A study by Chan et al. (2016) also found that there had been talks about relocating an entire Hani village out of the protected zone, but the plans were temporarily stalled, as authorities could not find a suitable alternative site for the community.

Heritage preservation, land grabbing and resettlement in the Angkor Archaeological Park, Cambodia

The Angkor Archaeological Park (AAP) is situated in Cambodia's Siem Reap Province on the northern shores of the Tonle Sap Lake, which is Southeast Asia's largest freshwater lake and has been described as the country's 'beating heart'. It is a place of exceptional archaeological significance, with magnificent monuments (Figure 8.1) such as Angkor Wat, Angkor Thom, Bayon and Ta Prohm, dating from the 9th to the 15th century when the Angkor region was the capital of an expansive Khmer empire (Gillespie, 2009). Preservation of the monuments for tourism consumption dates back to French colonialism when archaeologists and historians from the *École Française d'Extrême Orient* (EFEO) removed trees, monks and local communities from the expansive temple grounds to reinforce their own imagination of historic 'Indochina' and establish themselves as the sole interpreters of Angkorean and Khmer history (Winter, 2007).

Figure 8.1 Historic monuments at the Bayon Temple in Angkor Archaeological Park (AAP)

The Angkor Archaeological Park (AAP) was listed as a 'World Heritage in Danger' site by UNESCO in 1992, a year after the 1991 Paris Peace Accord, which brought relative political stability to post-conflict Cambodia. As the legislative, executive and judicial branches of Cambodia's transitional government were deemed weak and the entire cadastral infrastructure of the country had been destroyed under the genocidal regime of the Khmer Rouge (1975–1979), UNESCO authorities saw the urgent need of establishing legal and spatial boundaries to protect the cultural heritage sites against threats of land grabbing and unregulated construction (Winter, 2015). The AAP was subsequently subjected to a zoning process under the so-called Zoning and Environmental Management Plan (ZEMP), drawn up in 1993 and legally enforced by Royal Decree in 1994, with Zone 1 defined as a core zone with the highest level of protection and Zone 2 described as a buffer to protect the core zone (Gillespie, 2011; Winter, 2007). Regulations regarding these two zones have strong implications for the more than 120,000 residents divided into over 100 villages and hamlets (Figure 8.2), most of which had occupied the land prior to the AAP's World Heritage designation (Gillespie, 2009). Another three zones were established to designate protected cultural landscapes along rivers (Zone 3), additional sites of archaeological, anthropological and historic interest (Zone 4) and the socio-economic and cultural development zone of the wider Siem Reap region (Zone 5) (Miura, 2011a). Recognising the threats to the preservation of the AAP, UNESCO created the International Coordinating

Figure 8.2 Shops and street vendors in a village inside the core zone of the AAP

Committee for the Safeguarding and Development of Angkor (ICC), an administrative body co-chaired by France and Japan which were Cambodia's largest bilateral donor countries at the time, thereby further foreignising the legal and administrative infrastructure of the Angkor region (Winter, 2007).

Heeding calls for the establishment of a Cambodian-run management body to complement the ZEMP and ICC, a 1995 Law – further backed by a 1999 Law – assigned the management authority of the AAP to the *Autorité pour la Protection du Site et l'Aménagement de la Region d'Angkor* (APSARA), i.e. the Authority for the Protection and Management of Angkor and the Region of Siem Reap (Miura, 2011a; Winter, 2007). It also established the so-called 'Heritage Police' in 1997, under the authority of Cambodia's Ministry of Interior and trained by French police forces (Miura, 2011a). While the AAP is formally under the status of 'state public property', the customary land rights of residents in Zones 1 and 2 are also recognised. Villagers are entitled to remain in those zones as long as they do not expand their properties and do not sell their land to outsiders (see Box 8.1). However, in the early years, the zoning laws were not sufficiently made publicly known, and the boundaries remained largely unclear to many residents. This uncertainty was exploited by some Cambodian authorities – other than APSARA – to issue permits to build tourism facilities in the protected zones (Miura, 2011a). The national government also supported the expansion of hotels since the early 2000s, as its stated aim was to benefit economically from the commodification of the AAP for tourism purposes (Heikilla and Peycam, 2010). At the same time, the Heritage Police – often without consulting APSARA – imposed bans on many tradi- tional practices of local residents and collected fines from street vendors and rice farmers for allegedly illegal activities (Miura, 2011a).

Box 8.1 Rights of residents to their own land in Zones 1 and 2 of Angkor Archaeological Park

1   The villagers, who have homes and have lived in the Angkor Heritage Park for ages, can continue to live there without being forced to leave the village.
2   The villagers can demolish their old houses or build new ones with authorisation from APSARA Authority.
3   The villagers have the right to manage their own lands such as: trans- ferring land possession to their relatives – parents to children, or selling land to their neighbours to obtain money for living expenses. However, it is forbidden to buy and sell land for the purpose of making profitable business for companies or individuals to build hotels, restaurants, KTV (karaoke venues), etc.

Source: http://apsaraauthority.gov.kh/?page=front&lg=en

Many incidents of evictions, house demolitions and harassments have been reported in the context of the AAP in the early 2000s (e.g. Miura, 2011b). In 2017, more than 500 'illegally' constructed houses inside the AAP were demolished by police upon order of APSARA and pressure from UNESCO's Cambodia representative (Pech, 2017a, 2017b). APSARA argued that these buildings were hastily erected in the lead-up to Cambodia's communal elec- tions held in June 2017 and that their construction was unlawfully authorised by village and commune leaders (pers. comm., senior APSARA official, Feb- ruary 2019). In another case, one village in the vicinity of the AAP was asked to give up their rice fields in order to make way for a resettled community, but several households continue to resist this dispossession (Figures 8.3 and 8.4).

While forced evictions from the AAP have been infrequent in recent years, younger residents in particular are encouraged to move out of the two inner zones. In the late 2000s, APSARA created an 'eco-village for sustainable development' on an area of roughly 1,000 hectares in order to reduce the population pressure in the AAP (Kaliyann, 2013). The eco-village named Run Ta-Ek was designed by a Canadian architect and also features a scenic lake, a Buddhist pagoda and home- stay-style bungalows for tourists (Figures 8.5 and 8.6). APSARA provided free land (and free housing materials for the first 100 families) to entice inhabitants of the AHP to move to this area which is located a few kilometres off the northeastern corner of the Park's Zone 2. As of February 2019, only about 50 households were living in Run Ta-Ek, but there are plans to move an additional 800 families over the next few years (pers. comm., APSARA senior official, February 2019).

Figure 8.3 Community members guarding their rice fields against dispossession by APSARA

Figure 8.4 Police on patrol around the village facing dispossession of their rice fields

APSARA insists that the buildings are kept in a traditional way and that the land may never be sold to maintain a 'sense of permanence' (Miura, 2011b).

Meanwhile, land speculation in and around the AAP is booming. As land prices in Siem Reap Town are rapidly rising, tourism investors are increasingly looking for land closer to the AAP. A senior APSARA official expressed his frustration with these land grabs:

> Government authorities work only eight hours a day, but land investors work 24 hours a day, so how can we stop land speculation? Villagers are lured into selling their land due to the high land prices and some of them sell at night to avoid being fined. The commune leaders also do not respect their role but sign papers that allow outsiders to purchase land. Even the banks recognise such land as collateral. Cambodian people participate in the destruction of their own culture.
>
> (pers. comm., February 2019)

Prior to the COVID-19 pandemic, there were no signs that the tourism boom and the associated land speculation would subside any time soon. In 2018, record numbers of international visitors of the AAP were counted (2.59 million), a more than 5 per cent increase to the previous year (Cheng, 2019). The AAP also remains popular with domestic tourists who regard the monuments as both a symbol of national pride and a place of religious devotion.

Figure 8.5 Lake-side pagoda in Run Ta-Ek Eco-Village

Figure 8.6 A tourist bungalow in Run Ta-Ek Eco-Village

Gentrification in urban heritage sites: global overview and examples from South America

Tourism gentrification is a worldwide phenomenon and has particularly been noted in cities. It refers to the sudden and/or gradual displacement of original inhabitants that occurs when new touristic 'spaces of consumption' are created that increase land values and rents (Logan and Molotoch, 2007) and provide strong incentives for local governments and/or tourism investors to evict 'unwanted' urban residents or grab land from urban dwellers (Box 8.2).

Box 8.2 Gentrification

While there is no unified definition of 'gentrification', there is a consensus that it involves the displacement of a poorer or socially marginalised group by a wealthier stratum of society that also tends to exhibit different cultural characteristics. Gentrification can be driven by national and local governments through the change of zoning laws or simply the demolition of poorer or informal neighbourhoods that are deemed an 'eyesore', a security threat or a hindrance to modern urban development. Such state-led gentrification processes can be contrasted to a market-driven process which is characterised by capitalist accumulation of real estate property and the resulting increase of land/property prices and rents which makes housing unaffordable by the poorer classes. Most gentrification processes are marked by a combination of the two drivers.

Tourism gentrification can take three forms of displacement: (1) residential displacement, (2) commercial displacement, and (3) place-based displacement. The latter form refers to the loss of a sense of place and belonging experienced by residents when their customary spaces are increasingly invaded and consumed by tourists.

Source: Betancur, 2014; Bromley and Mackie, 2009; Gaffney, 2016;
Cocola-Gant, 2018

There is evidence that tourism gentrification is particularly prevalent in developing countries and emerging economies that rely on foreign visitors for economic growth and development (Cocola-Gant, 2018). Yet there are also numerous cases of tourism gentrification in the Global North. In Germany's capital Berlin, for instance, rising visitor numbers have contributed to widespread evictions and an increase in rents that made housing increasingly unaffordable for lower-income residents (Füller and Michel, 2014). In the southern US city New Orleans, tourism growth led to an escalation of property prices and the conversion of relatively affordable family homes into high-end condominiums, thereby pushing out poorer residents (Gladstone and Préau, 2008; Adams, van Hattum and English, 2009). Airbnb has also become a major factor of tourism gentrification by substituting long-term tenants with short-term visitors and by taking housing units off the real estate market, which has been empirically shown for the case of Barcelona, Spain (Cocola-Gant, 2018). In the Portuguese capital Lisbon tourism growth was regarded as a major solution to recover from the 2008 financial crisis, while the EU-enforced liberalisation of the housing market resulted in widespread housing rehabilitation efforts, whereby local residents were evicted for the opening of new hotels and short-term rentals for tourists (Cocola-Gant, 2018).

Many Latin American cities have instigated major projects of redeveloping their historical centres for cultural heritage tourism, often combined with sustained efforts of displacing classes of low social status (Betancur, 2014). In many instances, such projects have been a response to the expanding occupation of public spaces by informal traders which were seen as an obstacle to creating city centres that appeal to international tourists (Bromley and Mackie, 2009). Notable examples are Mexico City, Santiago de Chile, Quito and Lima, whereby international financial institutions (IFIs), such as the Inter-American Development Bank and the World Bank, often played a primary role by offering generous financial packages for these controversial projects. Yet even the Inter-American Development Bank's own studies concede that the "[p]reservation of urban historic centers generally involves gentrification, with higher-income residents and economic activities supplanting poorer ones" (Rojas, 1999, p. 16) and that such gentrification processes "expel low-income families and less profitable economic activities from the area. The poor lose access to cheap housing and to the economic and social opportunities offered by a downtown location" (ibid, p. 17). In several instances, the discourse employed by project implementers was to rescue the urban cultural heritage from lower classes and regain moral authority over historically important spaces, as in the case of Puebla in Mexico (Jones and Varley, 1999).

The case of La Boca, Buenos Aires, Argentina

La Boca is a southern neighbourhood in Argentina's capital Buenos Aires. Historically, this was a port of entry for Italian and other European immigrants and therefore emblematic for the formation of Argentina's urban working class (Rodríguez and Di Virgilio, 2016). It is also considered the birthplace of tango which put Buenos Aires on the world's cultural map and contributed to the city's self-proclaimed image as Latin America's 'cultural capital'. The first urban renewal project in La Boca was undertaken by the city government in 1993 around the construction of coastal flood-control measures with a US$120 million loan from the Inter-American Development Bank (Herzer, 2010). This was followed in the 2000s by further investments in neighbourhood beautification and tourism development (Rodríguez et al., 2011).

Until the late 1990s, tourism in La Boca was concentrated on the Caminito enclave as other areas were deemed unsafe for tourists due to the neighbourhood being one of the poorest in the city. Caminito is a walking street most closely associated with the merchandising of Tango culture, featuring colourful houses with art galleries, restaurants and street stalls (Janoschka and Sequera, 2016). After 2001, the municipal government tried to expand the tourism zone westwards by refurbishing historic façades, improving lighting and providing stronger police presence with the aim to include the many art studios that had previously been avoided by visitors due to security concerns (Wong, 2017). In 2012, legal provisions were enacted (Law N° 4353) to create a new arts district with 20 new cultural spaces by integrating sections of La Boca with its northern neighbour San Telmo, thereby establishing a cultural and touristic corridor that would link 180 sites of cultural consumption for visitors (Gobierno de la Cuidad de Buenos Aires, 2017).

This cultural megaproject – developed as a public-private partnership – triggered massive urban land speculation, forced evictions and homelessness (Sequera and Rodríguez, 2017). In 2016 alone, at least 1,100 residents were displaced, and 96 eviction trials were negotiated in city courts according to the International Tribunal of Evictions (2017). There were also reports about fire incidents in buildings, which were attributed to arson attacks allegedly instigated by urban investors (Rodríguez and Di Virgilio, 2016; Sequera and Rodríguez, 2017). More subtle approaches employed by local policy makers have involved fiscal exemptions for real-estate developers investing in the mystification and promotion of Tango culture, backed by UNESCO (Janoschka and Sequera, 2016).

In sum, heritage tourism development in Buenos Aires's La Boca and San Telmo districts has been marked by a combination of cultural and physical violence (Janoschka and Sequera, 2016). While tourism still remains in its enclave form and the poorer and underprivileged classes are being pushed from their original neighbourhood, the major winners of this process of 'forced gentrification' are Buenos Aires's middle and upper classes who can enjoy more diverse and much safer cultural spaces where they can socialise and entertain themselves alongside foreign tourists (Wong, 2017).

The case of Cuzco, Peru

Cuzco (or Cusco) is a major Peruvian city in the southern Andes, located at an altitude of 3,400 meters above sea level (Figure 8.7). Its proper name, 'Qosqo' in the Quechua language, means 'the navel of the world' which reflects its importance as the centre of the Inca Empire in the 15th century. Following the Spanish conquest in the 16th century, the urban structure developed by the Inca was mostly preserved, while baroque-style churches, monasteries and manor houses were built over the existing pre-Colombian buildings, turning Cuzco into a hybrid product of Inca imperial heritage and Hispanic colonial legacy (UNESCO, n.d.).

The 'City of Cuzco' obtained World Heritage status in 1983, together with 'The Historic Sanctuary of Machu Picchu' which most tourists that visit Cuzco have on their itinerary. The Ministry of Culture and the Provincial Municipality of Cuzco bear primary responsibility for the preservation of the city's cultural heritage programmes and the management of the World Heritage property and perform regular urban assessments, registration, protection, and control works. The Municipality has made extensive efforts towards

Figure 8.7 Panoramic view of the City of Cuzco in the Southern Andes region of Peru

Figure 8.8 The gentrified centre of the City of Cuzco as a 'safe tourist space'

creating a clean and safe environment for tourists (Steel, 2013). Cuzco's historic centre has been transformed from a residential zone into a huge commercial centre dedicated to international tourists who are able to pay the steep prices of the luxury hotels, restaurants and souvenir shops (Chion, 2009). While UNESCO regards new tourism development as a threat to the preservation and functional capacity of ancient buildings, Cuzco authorities have cleared the city of 'urban undesirables', such as beggars, street vendors and the urban poor, in order to provide tourists with a sense of comfort and security (Steel, 2013; Figure 8.8).

The first major 'social cleansing' of the city centre was instigated by former city mayor Carlos Valencia who considered informal trade and hillside slums as the most pressing planning issues of the city. A study by Bromley and Mackie (2009) found that more than 6,200 informal street vendors were forcibly displaced from central Cusco between 2001 and 2004, thereby freeing public spaces for middle-class residents and foreign and domestic tourists. A similar process of 'social cleansing' has been reported by Janoschka and Sequera (2016) who described how thousands of informal street vendors in the historic centre of Mexico City were first declared illegal and then evicted from public spaces in a military-style police raid in 2007.

This state-led gentrification has been accompanied by market-led gentrification, with external investors buying up prime property in the city's centre. A controversial section of Article 2, Law 29164, allows the allocation of tourism concessions for privately owned, four- to-five-star hotels and restaurant within national heritage sites, which caused intense protests by Cuzco's residents in

2008 (Knight et al., 2017). As local investors are increasingly unable to buy property in the historic centre and local residents are unable to afford the excessively high rents in central parts of the city, impoverished quarters in the city's outskirts are rapidly expanding. Displaced urban residents are joined in these neighbourhoods by scores of rural migrants who came to Cuzco in the hope of finding new livelihood opportunities in the tourism sector, only to be confronted with the harsh reality of being offered low-paid, precarious jobs or finding no employment at all (Steel, 2013). In these crowded informal settlements, access to water, electricity, public transport as well as sanitary and health services is extremely limited, which contrasts starkly with the amenities that are provided for globally connected tourists in Cuzco's historic centre.

The growth of cultural tourism in Cuzco has also important implications for the peri-urban fringes beyond the city's hillside squatter settlements. For several decades, the local tourism sector has campaigned for building a new international airport in Chinchero, about 30 kilometres north of Cuzco. The Peruvian government approved the project in 2012 and started expropriating Indigenous land for the airport project in 2013 (Garcia, 2020). Opponents of the controversial infrastructure project claim that it would not only have devastating impacts on Indigenous communal land rights but destroy the heartland of Inca civilisation (Collyns, 2019).

Concluding remarks

While tourism scholars and practitioners tend to assign positive values to visits of cultural heritage as "a shared common good by which everyone benefits" (Silverman and Ruggles, 2007, p. 3), the examples in this chapter provide evidence of the contentious entanglement of heritage tourism with universalist cultural ideologies, state power and localised abuse of human rights. The case of Guatemala's Maya Biosphere Reserve shows how contrasting visions about tourism, heritage conservation and environmental protection can lead to entrenched conflicts between communities and other actors, in this case archaeologists, corporations, drug barons, the national government and politicians from a foreign power. The case studies from China demonstrate how the designation of World Heritage Sites and associated tourism developments have led to restrictions on traditional livelihood activities and the resettlement of populations, resulting in protracted tensions between the government, corporate actors and local communities. The case of the Angkor Archaeological Park in Cambodia testifies to the colonial legacies of cultural heritage management and shows how the interpretations and regulations imposed by non-Cambodians (e.g. UNESCO, bilateral donors, Western archaeologists) marginalises contemporary Cambodian culture, silences local voices, and infringes on local residents' rights to property and adequate housing. The inherent conflict between universal heritage and local rights is also evidenced in the cases of Buenos Aires' La Boca district and the ancient Inca capital Cuzco where a combination of violent state-led gentrification and market forces has evicted residents and informal vendors from the city centres to the urban fringes. Their livelihoods have become increasingly precarious due to the lack of basic amenities and economic opportunities.

Such processes of gentrification are further examined in the next chapter which discusses how sports mega-events and large-scale tourism infrastructure projects for airports and railways have been used to legitimise urban 'beautification' schemes, forced land acquisitions and involuntary resettlements. Examples from South Africa, Brazil, India, Laos and Mexico will show the destructive impacts such processes have had on local livelihoods and cultures.

References

Adams, V., van Hattum, T. and English, D. (2009) 'Chronic disaster syndrome: Displacement, disaster capitalism, and the eviction of the poor from New Orleans', *American Ethnologist*, 36 (4): 615–636.

Ashworth, G., Graham, B. and Tunbridge, J. (2007) *Pluralising Pasts: Heritage, Identity and Place in Multicultural Societies*. Pluto: London.

Betancur, J.J. (2014) 'Gentrification in Latin America: Overview and critical analysis', *Urban Studies Research*, 4: 1–14.

Bocci, C., Fortman, L., Sohngen, B. and Millian, B. (2018) 'The impact of community forest concessions on income: an analysis of communities in the Maya Biosphere Reserve', *World Development*, 107: 10–21.

Bromley, R.D.F. and Mackie, P.K. (2009) 'Displacement and the new spaces for informal trade in the Latin American city centre', *Urban Studies*, 46 (7): 1485–1506.

Chan, J.H., Iankova, K., Zhang, Y., McDonald, T. and Qi, X. (2016) 'The role of self-gentrification in sustainable tourism: Indigenous entrepreneurship at Honghe Hani Rice Terraces World Heritage Site, China', *Journal of Sustainable Tourism*, 24 (8–9): 1262–1279.

Chion, M. (2009) 'Cambios en el centro histórico del Cuzco: espacios turísticos, espacios culturales [Changes in the historic centre of Cuzco: tourist spaces, cultural spaces]' in Nieto Degregori, L. (ed.) *El ombligo se pone piercing: Identidad, patrimonio y cambios en al Cuzco* [The Navel Gets Pierced: Identity, Heritage and Changes in Cuzco]. Centro Guaman Poma de Ayala: Lima, pp. 115–150.

Cocola-Gant, A. (2018) 'Tourism gentrification' in Lees, L. and Phillips, M. (eds) *Handbook of Gentrification Studies*. Edward Elgar Publishing: Cheltenham, Northampton, pp. 281–293.

Devine, J. (2014) 'Counterinsurgency ecotourism in Guatemala's Maya Biosphere Reserve', *Environment and Planning D: Society and Space*, 32: 984–1001.

Devine, J. (2017) 'Colonizing space and commodifying place: Tourism's violent geographies', *Journal of Sustainable Tourism*, 25 (5): 634–650.

Devine, J. and Ojeda, D. (2017) 'Violence and dispossession in tourism development: A critical geographical approach', *Journal of Sustainable Tourism*, 25 (5): 605–617.

Devine, J.A., Wrathall, D., Currit, N., Tellman, B. and Langarica, Y.R. (2020) 'Narco-cattle ranching in political forests', *Antipode*, 52 (4): 1018–1038.

Füller, H. and Michel, B. (2014) '"Stop being a tourist!" New dynamics of urban tourism in Berlin-Kreuzberg', *International Journal of Urban and Regional Research*, 38 (4): 1304–1318.

Gaffney, C. (2016) 'Gentrifications in pre-Olympic Rio de Janeiro', *Urban Geography*, 37: 1132–1153.

Gao, J., Lin, H. and Zhang, C. (2020) 'Locally situated rights and the "doing" of responsibility for heritage conservation and tourism development at the cultural landscape of Honghe Hani Rice Terraces, China', *Journal of Sustainable Tourism*. doi:10.1080/09669582.2020.1727912.

Garcia, P. (2020) 'Indigeneity in the air: The case of Chinchero airport in Cusco, Peru', *Bulletin of Latin American Research*, 39 (2): 157–171.

Gillespie, J. (2009) 'Protecting world heritage: Regulating ownership and land use at Angkor Archaeological Park, Cambodia', *International Journal of Heritage Studies*, 15 (4): 338–354.

Gillespie, J. (2011) 'Legal pluralism and world heritage management at Angkor, Cambodia', *Asia-Pacific Journal of Environmental Law*, 14 (1&2): 1–19.

Gladstone, D. and Préau, J. (2008) 'Gentrification in tourism cities: evidence from New Orleans before and after Hurricane Katrina', *Housing Policy Debate*, 19 (1): 137–175.

Gobierno de la Ciudad Autónoma de Buenos Aires (2017) *Transformación del Distrito* [Transformation of the District]. Available at: www.buenosaires.gob.ar/distritodelasa rtes/transformacion-deldistrito (accessed 16 February 2019).

Herzer, H. (2010) *Con el corazon mirando al sur: Transformaciones en el sur de la ciudad de Buenos Aires* [With the Heart Looking Southward: Transformations in the South of the City of Buenos Aires]. Espacio Editorial: Buenos Aires.

Heikilla, E.J. and Peycam, P. (2010) 'Economic development in the shadow of Angkor Wat: Meaning, legitimation, and myth', *Journal of Planning Education and Research*, 29 (3): 294–309.

International Tribunal of Evictions (2017) Case from Argentina, Buenos Aires: Evictions and gentrification in the historic and tourist neighbourhood La Boca. Available at: www.tribunal-evictions.org/international_tribunal_on_evictions/evictions_cases/6th_ session_2017/case_from_argentina_buenos_aires_evictions_and_gentrification_in_the _historic_and_tourist_neighbourhood_la_boca (accessed 16 February 2019).

IWGIA and FPP (2015) Promotion and protection of the rights of indigenous peoples with respect to their cultural heritage in the context of the implementation of UNESCO's World Heritage Convention. Available at: www.ohchr.org/Docum ents/Issues/IPeoples/EMRIP/CulturalHeritage/IWGIA.pdf (accessed 12 November 2018).

Janoschka, M. and Sequera, J. (2016) 'Gentrification in Latin America: Addressing the politics and geographies of displacement', *Urban Geography*, 37 (8): 1175–1194.

Jones, G.A. and Varley, A. (1999) 'The reconquest of the historic centre: urban conservation and gentrification in Puebla, Mexico', *Environment and Planning A*, 31 (9): 1547–1566.

Knight, D.W., Cottrell, S.P., Pickering, K., Bohren, L. and Bright, A. (2017) 'Tourism-based development in Cusco, Peru: Comparing national discourses with local realities', *Journal of Sustainable Tourism*, 25 (3): 344–361.

Li, Y., Lau, C. and Su, P. (2020) 'Heritage tourism stakeholder conflict: A case of a World Heritage Site in China', *Journal of Tourism and Cultural Change*, 18 (3): 267–287.

Logan, J. and Molotoch, H. (2007) *Urban Fortunes: The Political Economy of Place* (2nd edition). University of California: Berkeley, Los Angeles.

Luciano, P.A. (2017) 'Where are the edges of a protected area? Political dispossession in Machu Picchu, Peru', *Conservation and Society*, 9 (10): 35–41.

Lunstrum, E. and Ybarra, M. (2018) 'Deploying difference: security threat narratives and state displacement from protected areas', *Conservation and Society*, 16 (2): 114–124.

Miura, K. (2011a) 'World heritage making in Angkor: Global, regional, national and local actors, interplays and implications' in Hauser-Schäublin, B. (ed.) *World Heritage Angkor and Beyond: Circumstances and Implications of UNESCO Listings in Cambodia.* Göttingen University Press: Göttingen, pp. 9–31.

Miura, K. (2011b) 'Sustainable development in Angkor: Conservation regime of the old villagescape and development' in Hauser-Schäublin, B. (ed.) *World Heritage Angkor and Beyond: Circumstances and Implications of UNESCO Listings in Cambodia.* Göttingen University Press: Göttingen, pp. 121–144.

Mowforth, M. and Munt, I. (2016) *Tourism and Sustainability: Development, Globalisation and New Tourism in the Third World* (4th edition). Routledge: London, New York.

Rahder (2015) 'But is it a basin? Science, controversy, and conspiracy in the fight for Mirador, Guatemala', *Science as Culture*, 24 (3): 299–324.

Rodríguez, M.C. and Di Virgilio, M.M. (2016) 'A city for all? Public policy and resistance to gentrification in the southern neighborhoods of Buenos Aires', *Urban Geography*, 37 (8): 1215–1234.

Rodríguez, M.C., Mejica, S.A., Rodríguez, M.F., Gómez Schettini, M. and Zapata, M.C. (2011) 'La política urbana "PRO": Continuidades y cambios en contextos de renovación en la ciudad de Buenos Aires [The urban policy "PRO": continuities and changes in the contexts of renewal in the city of Buenos Aires]', *Cuaderno Urbano*, 1 (1): 81–101.

Rojas, E. (1999) *Old Cities, New Assets: Preserving Latin America's Urban Heritage.* Inter-American Development Bank: Washington, DC.

Sequera, J. and Rodríguez, T. (2017) 'Turismo, abandono y desplazamiento: Mapeando el barrio de La Boca en Buenos Aires [Tourism, abandonment and displacement: Mapping the La Boca neighbourhood in Buenos Aires]', *Journal of Latin American Geography*, 16 (1): 117–137.

Silverman, H. and Ruggles, D.F. (2007) 'Cultural heritage and human rights' in Silverman, H. and Ruggles, D.F. (eds.) *Cultural Heritage and Human Rights.* Springer: New York, pp. 3–22.

Steel, G. (2013) 'Mining and tourism: Urban transformations in the intermediate cities of Cajamarca and Cusco, Peru', *Latin American Perspectives*, 40 (2): 237–249.

Su, M.M., Wall, G. and Xu, K. (2016) 'Heritage tourism and livelihood sustainability of a resettled rural community: Mount Sanqingshan World Heritage Site, China', *Journal of Sustainable Tourism*, 24 (5): 735–757.

Timothy, D.J. (2015) 'Cultural heritage, tourism and socio-economic development' in Sharpley, R. and Telfer, D.J. (eds) *Tourism and Development: Concepts and Issues* (2nd edition). Channel View Publications: Bristol, Buffalo, Toronto, pp. 237–249.

US Congress (2019) *S.3131 - Mirador-Calakmul Basin Maya Security and Conservation Partnership Act of 2019.* Available at: www.congress.gov/bill/116th-congress/sena te-bill/3131/text (accessed 10 July 2020).

Wang, L. and Yotsumoto, Y. (2019) 'Conflict in tourism development in rural China', *Tourism Management*, 70: 188–200.

Winter, T. (2007) *Post-Conflict Heritage, Postcolonial Tourism: Culture, Politics and Development at Angkor.* Routledge: London, New York.

Wong, L.K.W. (2017) *Urban Transformations in Latin America's Cultural Capital: An Analysis of City Branding and its Impact on Gentrification in Buenos Aires.* Masters Thesis in Latin American Studies, Leiden University, Leiden.

Zhang, C., Fyall, A. and Zheng, Y. (2015) 'Heritage and tourism conflict within world heritage sites in China: A longitudinal study', *Current Issues in Tourism*, 18 (2): 110–136.

Media sources/websites

Cheng, S. (2019) 'Angkor hosts 2.6M visitors', *Phnom Penh Post*, 2 January. Available at: www.phnompenhpost.com/business/angkor-hosts-26m-visitors (accessed 18 February 2019).

Collyns, D. (2019) '"It would destroy it": new international airport for Machu Picchu sparks outrage', *The Guardian*, 15 May. Available at: www.theguardian.com/cities/20 19/may/15/archaeologists-outraged-over-plans-for-machu-picchu-airport-chinchero (accessed 12 July 2020).

Cuffe, S. (2016) 'Controversial park plans in Guatemala's Maya Biosphere Reserve', *Mongobay*, 17 June. Available at: https://news.mongabay.com/2016/06/controversia l-park-plans-in-guatemalas-maya-biosphere-reserve/ (Accessed on 9 February 2019).

García, B. (2020) 'The suspicious initiative of an archaeologist to "save" an ancient Mayan city', *Al Día*, 18 June. Available at: https://aldianews.com/articles/culture/ social/suspicious-initiative-archaeologist-save-ancient-mayan-city/58920   (accessed 12 July 2020).

Kaliyann, T. (2013) 'Eco village attracts homestay tourist dollars', *Phnom Penh Post*, 12 August. Available at: www.phnompenhpost.com/siem-reap-insider/eco-village-attra cts-homestay-tourist-dollars (accessed 18 February 2019).

Maya Biosphere Watch (2020) 'U.S. senators draft legislation that infringes on land rights in Guatemala and Mexico', 4 January. Available at: https://mayabiospherewatc h.com/u-s-senators-draft-legislation-that-infringes-on-land-rights-in-guatemala-and-mexico/ (accessed 12 July 2020).

Pearce, F. (2020) 'Parks vs. people: In Guatemala, communities take best care of the forest', *Yale Environment 360*, 18 June. Available at: https://e360.yale.edu/features/pa rks-vs-people-in-guatemala-communities-take-best-care-of-the-forest (accessed 12 July 2020).

Pech, S. (2017a) 'Forces demolish illegal homes in Angkor park', *Khmer Times*, 20 June. Available at: www.khmertimeskh.com/news/39476/forces-demolish-illegal-homes-i n-angkor-park/ (accessed 18 February 2019).

Pech, S. (2017b) 'Illegal building at Angkor site must go, says Unesco', *Khmer Times*, 27 June. Available at: www.khmertimeskh.com/news/39692/illegal-building-at-a ngkor-site-must-go–says-unesco/ (accessed 18 February 2019).

UNESCO (n.d.) 'Biosphere Reserves – Learning Sites for Sustainable Development'. Available at: www.unesco.org/new/en/natural-sciences/environment/ecological-sci ences/biosphere-reserves/ (accessed 12 February 2019).

UNESCO (n.d.) 'City of Cuzco'. Available at: https://whc.unesco.org/en/list/273 (accessed 12 February 2019).

9 The displacement effects of sports mega-events and large-scale tourism infrastructure development

International sports mega-events, such as the FIFA Men's World Cup, the Commonwealth Games and the Olympic Games, attract hundreds of thousands of spectators and generate increased economic investments and tourism revenues (Maharaj, 2015). The scale of such events requires the allocation of large amounts of scarce public resources and entails profound transformations of the urban built environment, thereby affecting the lives and livelihoods of many urban residents, particularly those at the fringes of society. The 21st century has seen an increasing shift of the global geography of sport mega-event hosting to countries of the Global South, raising profound questions around organisational implications, securitisation and social exclusion in political and societal settings that are marked by extreme income polarities, uneven power relations and disputed state agendas (Gaffney, 2010; Cornelissen, 2011).

Similarly, large-scale development of tourism infrastructure, such as airport and railway constructions and upgrades can have profound impacts on local people through compulsory land acquisitions, forced displacement, and insufficient compensation for loss of farmland, residential land and communal resources (Sims, 2015; Mteki, Murayama and Nishikizawa, 2017; Camargo and Vázquez-Maguirre, 2020). Indigenous peoples, ethnic minority groups, informal settlers and other marginalised groups are particularly vulnerable to the various social, cultural, economic and environmental injustices that tend to be associated with such projects that are discursively justified as being in the interest of the wider public.

In the following section, the adverse impacts of sports mega-events on the land and housing rights of poor and marginalised groups are examined, drawing on a brief global overview and more detailed analysis of recent mega-events in South Africa and Brazil. Then the physical and economic displacement effects of airport constructions and expansions in India and Laos are explored. The final section examines the discourses and conflicts around the controversial 'Mayan Train' mega-project on the Yucatán Peninsula in southern Mexico.

Sports mega-events: the darker side of the global spectacle

International sports mega-events have become an important force in global urban development and are often associated with mass urban renewal schemes and large-scale displacement of people (COHRE, 2007; Porter, 2009; Davis, 2011; Müller, 2015; Müller and Gaffney, 2018). Displacement of urban residents through such mega-events is not a new phenomenon and certainly not confined to countries in the Global South. One of the earlier and probably most notorious examples is Nazi Germany which hosted the Olympic Games in 1936 and cleared Berlin of its homeless people and informal residents to provide a 'clean' image to the world (Newton, 2009). In the 1990s, the City of Atlanta deported about 25,000 homeless people and closed most of the city's facilities for the homeless ahead of the 1996 Olympic Games (Warrall, 2016). The London Olympics in 2012 were associated with the forced removals of several hundred low-income households and small businesses (Raco and Tunney, 2010; Watt, 2013).

Emerging economies in Asia displaced hundreds of thousands of urban residents in the context of sports mega-events, although these were not the only drivers at the time. 700,000 residents in the South Korean capital Seoul were relocated in the run-up to the 1988 Olympic Games and more than a million residents had to make way for the 2008 Beijing Olympic Games as part of larger urban redevelopment strategies (Porter, 2009; Müller, 2015). About 200,000 residents were evicted from Delhi's vast informal settlements, about 35,000 families were removed from public property and over 300,000 street vendors lost their livelihoods in preparation for the 2010 Commonwealth Games (Maharaj, 2015).

In addition to forced removal of urban residents, i.e. through expropriation or violent evictions by state forces, population 'redistribution' in the context of sports mega-events can also occur through market mechanisms in the absence of public policies that guarantee the right to affordable housing (Müller and Gaffney, 2018). Lenskyj (2002) reported how the 2000 Sydney Olympics exacerbated the housing gap and intensified homelessness and housing problems through changes in property values and urban land grabbing (cited in Newton, 2009).

The upcoming Tokyo Olympics – recently postponed to 2021 due to the COVID-19 pandemic – are a stark reminder that mega-events organised in the Global North can have significant social and ecological ripple effects on the Global South. The Rainforest Action Network (RAN) asserts that significant volumes of rainforest wood have been used to build Olympic facilities, including Tokyo's new National Olympic Stadium (RAN, n.d.). The timber was ostensibly supplied by a Malaysian logging company with a proven history of rainforest destruction, illegal logging and human rights abuses. Together with 46 other civil society organisations, RAN delivered an open letter to the International Olympic Committee (IOC) and Tokyo 2020 Olympic authorities, at the start of the IOC Executive Board Meeting in Lima, Peru, in September 2017. The letter reiterated concerns about the legitimacy and accountability of the IOC's sustainability commitments and the reputation and credibility of the Olympic Games. It further criticised the Olympics "for knowingly exploiting tropical forests and potentially fueling human rights abuses in the construction and implementation of the games". Finally, it urged Olympic authorities "to adopt robust social and environmental safeguards or face further criticism for fueling rainforest destruction, illegal logging and human rights violations". RAN also accuses one of the major sponsors of the Tokyo Olympics, Nissin Food, of not taking sufficient measures to avoid the use of so-called "Conflict Palm Oil" in its food products, such as cup noodles. The NGO urges the company "to strengthen its commitment to protect rainforests and peatlands and uphold human rights including land rights of Indigenous and local community and labor rights" (RAN, n.d.).

Sports mega-events in low-income countries with high incidence of poverty have also triggered land grabs and dispossession. When Laos hosted the Southeast Asian Games in 2009, much of the infrastructure was built with financial and technical support from the Chinese and Vietnamese governments. The debt of a US$80 million loan from the China Development Bank for building the national stadium was paid through a 300-hectare land concession for a special economic zone in a biodiversity-rich marshland in the outskirts of the Lao capital Vientiane (Stuart-Fox, 2009; Boliek, 2016). The That Luang Marsh did not only provide valuable agricultural land to its former beneficiaries but was also of crucial importance for flood control and natural wastewater treatment services in the capital (Schuettler, 2008). A study by Gerrard (2004) calculated that the goods and services provided by the marshland had amounted to nearly US$5 million annually, with 40 per cent of the benefits accruing to local residents. Hundreds of residents initially refused to make way for the US$1.6 billion commercial and tourism development project implemented by the Shanghai Wan Feng Group, as they considered the compensation offered as far too low (Vandenbrink, 2013). As a result, the Chinese real estate company had to postpone its plans to complete the urban development project ahead of the Asia-Europe Meeting (ASEM) forum in November in 2012 (Economist Intelligence Unit, 2012).

Even ten years later, the long-term impacts of the land-for-debt swap reverberate in Lao PDR's capital. In 2019 and 2020, hundreds of residents from several peri-urban villages protested against compulsory land acquisitions for a 13km expressway built by a Lao-Chinese consortium that will connect the That Luang Marsh special economic zone to another suburb of Vientiane (Gerin, 2019; Finney, 2020). The infrastructure investments for the Southeast Asian Games had also ripple effects in other provinces. A similar 'land-for-capital' deal as for the That Luang Marsh area was negotiated with a Vietnamese real estate company that contributed to building the US$19 million athlete's village for the Games and was granted a 90-year land concession of 10,000 hectares in one of Laos' southern provinces (Fuller, 2009).

Over the past decade, displacement through sports mega-events was most pronounced for the 2010 World Cup in South Africa, the 2014 FIFA World Cup in Brazil and 2016 Rio Olympic Games (Cummings, 2015; Müller & Gaffney, 2018). These cases are discussed in more detail hereafter.

The 2010 FIFA World Cup in South Africa: rendering poverty invisible by demolishing and upgrading informal settlements

The 2010 FIFA World Cup was proclaimed by the South African government as a motor of economic growth and an opportunity to change the African continent's persistent international image as a region of poverty and crisis (Steinbrink et al., 2011). South Africa spent about US$3 billion hosting the World Cup, slightly less than Germany which hosted this prestigious football event four years earlier but carried a relatively lower financial burden in relation to its much higher gross domestic product. An additional challenge for South Africa in organising this mega-event was that its ranking as the world's most unequal country in terms of income distribution. South Africa has also been plagued by a persistent housing crisis, where millions of people continue to live in informal squatter settlements.

In the run-up to the World Cup, many informal settlements in several of the host cities were demolished and extensive upgrading processes were instigated to 'beautify' the urban centres for the much-anticipated event (Steinbrink, Haferburg and Ley, 2011; Maharaj, 2015). A major flagship project that coincided with the approval of South Africa's bid for the 2010 World Cup by the FIFA was the N2 Gateway project in Cape Town, which had the objective to redevelop a 10km long squatter strip along the N2 motorway connecting the international airport to the city centre (Newton, 2009). Thousands of residents were relocated from these informal settlements, including the *Joe Slovo Settlement*, to so-called 'temporary relocation areas' in the impoverished city outskirts, which have been described as overcrowded refugee-style camps without electricity and only shared sanitary facilities (Steinbrink et al., 2011; Maharaj, 2015). While most relocated residents were initially given the prospect to return to a redeveloped settlement and be provided with subsidised housing, it later emerged that there were not enough housing units and that the majority could not afford the high rents in the newly built houses (COHRE, 2009).

Johannesburg – another 2010 World Cup venue and the city with the highest income inequality in the world – also saw major evictions and 'upgrading' projects. A large-scale urban renewal project around the Ellis Park Stadium, the so-called 'Greater Ellis Park Development Plan', excluded affected residents from any decision-making process and resulted in the forced eviction of many of the area's former inhabitants (Steinbrink, Haferburg and Ley, 2011). These processes went along with a major effort of 'securitising' this mega-event (Cornelissen, 2011). In compliance with FIFA requirements, hundreds of informal traders or street vendors were expelled from the 'exclusion zones' of the two stadiums, Soccer City and Ellis Park (Steinbrink, Haferburg and Ley, 2011). Such forms of economic displacement were also recorded from other venues. Altogether, it was estimated that about 100,000 mostly female street vendors temporarily lost their livelihoods during the 2010 FIFA World Cup (Maharaj, 2015).

The 2014 FIFA World Cup and the 2016 Rio Olympics in Brazil: exclusion games, forced evictions and state-led gentrification

Similar to the case of South Africa and the 2010 FIFA World Cup, Brazil regarded its successful bids for the 2014 FIFA World Cup and the 2016 Summer Olympic Games as a major opportunity to showcase an emerging powerhouse (Brazil) and a world-class global metropolis (Rio de Janeiro) to the rest of the world. Earlier, Brazil had hosted the 2007 Pan American Games which resulted in the removal of informal settlements (*fávelas*) around the athletes' housing complex (Gaffney, 2016). Like in South Africa, these mega-events primarily benefitted the urban elites and exacerbated the plight of the poor and marginalised groups that suffered from forced displacements, loss of livelihoods and serious violations of human rights. One study maintains – based on official data provided by the city administration – that more than 22,000 families or approximately 77,000 people were relocated in the city of Rio de Janeiro during the period 2009–2015 (Comitê Popular da Copa e das Olimpíadas do Rio de Janeiro, 2015). Strategies employed by city administrators ranged from selective eviction of *fávela* residents to relocation of inhabitants to social housing complexes in the periphery of the municipality (Janoschka and Sequera, 2016). In almost half of these cases, the official discourse was that these communities were 'at risk' from natural hazards, such as flooding and landslides (McGuirk, 2016).

Projects of *fávela remoção* (removal of informal settlements) are not a recent phenomenon in Rio de Janeiro. In the 19th century, municipal planners evicted thousands of informal settlers from the city centre with the ambition to turn it into a 'Tropical Paris', with a second wave of evictions following in the 1960s. When Rio hosted the Pan American Games in 2007, several partial favela removals were executed around venues in the suburb of Barra da Tijuca, hailed by the then president of the Brazilian Association of Travel Agents who cynically argued that "without the favelas blighting the landscape, tourism levels would rise, the profits of which could be channelled into fighting poverty" (Phillips, 2005, n.p.).

In 2008, a few months after winning the bid for the 2016 Olympic Summer Games, the state government of Rio de Janeiro instigated a highly controversial programme of sending so-called *Unidades Pacificadoros Policiais* (UPPs) or Police Pacification Units into dozens of *fávelas* within the Olympic clusters, in a co-funding arrangement with private investors (Gaffney, 2016; Janoschka and Sequera, 2016). The obvious aim of this measure was to 'pacify' unruly parts of the urban population and establish complete control over urban spaces that were deemed essential for a smooth organisation of the upcoming mega-event. One of the side effects of the UPP programme was the opening of backpacker hostels, nightclubs and adventure tours in these 'pacified *fávelas*', which intensified gentrification pressures and instigated conflicts between original inhabitants and developers (Gaffney, 2016).

The preparations for the 2016 Rio Olympics entailed profound changes to the mega-city's infrastructure, including the building of two Olympic parks and the Olympic village, the construction of four new rapid transit bus lines, and an extension of the metro (Müller and Gaffney, 2018). Several *fávelas* were demolished to build this expansive Olympic infrastructure and thousands of favela inhabitants were reportedly killed (Comitê Popular da Copa e das Olimpíadas do Rio de Janeiro, 2014). Even settlements where residents had formal land entitlements were not safe from the bulldozers; Vila Autódromo, a settlement of several hundred families with long-term leases granted earlier by the state government of Rio de Janeiro, lost 97 per cent of its residents to the construction of the International Broadcasting Centre and associated parking lots and road works (Silvestre and de Oliveira, 2012; Müller and Gaffney, 2018). Only 20 families that succeeded in resisting eviction were eventually accorded the right to stay in their community in August 2018 as part of the first-ever collectively negotiated rehousing agreement in Rio de Janeiro (Talbot, 2018).

Other communities that were facing eviction lived in proximity to the Tom Jobim International Airport which underwent a major expansion ahead of the 2016 Rio Olympics. Thousands of families had earlier been evicted from sites of airport upgrading and expansion projects in the 2014 FIFA World Cup venues Porto Alegre and Curitiba (Maharaj, 2015). This brings us to the issue of large-scale tourism infrastructure development discussed hereafter.

Opening the gates for mass tourism: airport constructions in India and Laos

In a 2016 press release, the International Air Transport Association (IATA) announced that it expected 7.2 billion passengers to travel in 2035, a near doubling of the 3.8 billion air travellers in 2016 (IATA, 2016). According to the IATA's projection, the five fastest growing markets in terms of additional passengers per year over the forecast period would be China, the United States, India, Indonesia and Vietnam. It was expected that much of this growth would be driven by tourism. It remains to be seen to what extent these forecasts will still hold, given that the ongoing COVID-19 pandemic has caused major disruptions to international travel – and to air traffic in particular – and that it may take several years for the global airline industry to recover from this downturn.

In December 2018, the Chinese government announced that the country will build 216 new airports by 2035, bringing the total number of airports in the country to 450 (*Tourism News Live*, 2018). Earlier in 2018, India's State Minister for Civil Aviation had stated in an interview that India plans to build 100 more airports for one billion flyers by 2035, with most of the US$60 billion financial capital needed to be provided by the private sector (Kuronuma, 2018). These massive infrastructure projects are expected to be associated with substantial environmental and social costs, including the forced relocation of thousands of people from the new airport construction sites.

Controversial greenfield airport development in Goa, India

In India, greenfield airports play a significant role in accelerating the urbanisation of the country and increasing domestic and international tourist numbers (Dalei and Singh, 2015). The Indian federal government established a national Greenfield Airport Policy in 2008, when it put in place a more liberalised approach to airport construction and promoted private participation and public–private partnerships (PPPs) in greenfield airport development (Nielsen, 2017). Yet searching for suitable sites for greenfield airports poses particular challenges in a country with a population density of 382 people per km^2, more than eight times the global average. On paper, land acquisition for the greenfield airport projects has become more difficult in the light of the new land acquisition act (Dalei and Singh, 2015; cf. Neef and Singer, 2015). Yet, in fact, infrastructure projects that are in the interest of national development (which includes growth within the tourism sector) are being fast-tracked under the government of Prime Minister Narendra Modi (see Box 9.1).

Box 9.1 Greenfield development and land acquisition by the state in India

Greenfield development refers to infrastructural developments in areas previously used for agriculture, forestry or rural amenity conservation. It evokes images of emptiness and opportunity and fosters visions of starting large infrastructure projects on 'virgin' lands. Yet, in reality, very few such projects can be built entirely on empty lands or in political or historical vacuums. Rather, greenfield project developers have to deal with "social and natural environments that are already inscribed with human and non-human history, habitation and activity" (Nielsen, 2017, p. 844).

In colonial India, the category of 'wasteland' had emerged in the 18th century as a key discursive tool through which 'idle' or 'unproductive' land was refashioned as 'untapped land', i.e. land that has yet to be tapped for its commercial potential (Bennike, 2017). This category was later adopted by the post-colonial state and employed as a land classification technique (Baka, 2013). Classifying land as 'wasteland' makes it much easier for the Indian federal government, local administrations and private investors to justify their land acquisition for major infrastructure project. This logic of capturing the potential of ostensibly marginal, virgin and idle land ties in well with the global land grab discourse (Neef, 2014; Nielsen, 2017).

In most cases, greenfield developments and other large infrastructure projects in India involve *land acquisitions by the state*. Until 2013, land acquisition and takeovers of land by the state for "public purpose" invoked the principle of "eminent domain" and was governed by a legal instrument from colonial times, the Land Acquisition Act of 1894. It was replaced by the Right to Fair Compensation and Transparency in Land Acquisition, Rehabilitation and Resettlement Act 2013 (LARR, 2013), which came into effect on 1 January 2014. The act introduced a number of safeguard mechanisms for farmers at risk of displacement and more generous compensation and resettlement packages for the displaced. However, amendments to the act under the Modi government have watered down some of its central clauses (Neef and Singer, 2015). The amendments include "a fast track process for defence and defence production, rural infrastructure including electrification, housing for poor including affordable housing, industrial corridors and infrastructure projects" (Government of India, 2014, n.p.).

In 2014, Modi launched the 'Make in India' programme which promoted India as the world's prime greenfield foreign direct investment (FDI) destination for 2015 and allows 100 per cent FDI for greenfield airport projects (Nielsen, 2017). In the same year, it was announced that 10–15 greenfield airports were in the planning stage (Firstpost, 2014). Only one year later, 14 greenfield airports received in-principle approval from the Indian government, with the planned Mopa Airport in Goa requiring the second highest investment (Sen, 2015). The 'Make in India' programme encourages Indian airports to adopt the special economic zone (SEZ) 'Aerotropolis model', whereby an airport is at the centre of surrounding facilities, such as luxury hotels, shopping and entertainment facilities, convention, trade and exhibition complexes, golf courses, and sport stadiums (Nielsen, 2017). All greenfield airport developments in India include processes of enclosure, eviction and environmental transformation. The largest greenfield airport project, the Navi Mumbai Airport, expected to be fully operational in 2023, required the displacement of around 2,700 families from ten villages and the destruction of ecologically important mangrove areas.

Another major greenfield project has been planned since 1999 in Goa, the smallest and richest state of India. Goa's primary industry is tourism, with the state accounting for 12 per cent of international tourist arrivals in India (Government of Goa, 2015). According to Goa's Department of Tourism, the state welcomed nearly 5.3 million visitors in 2015, double the number of tourist arrivals in 2010. Yet tourism development in Goa has also been met with fierce local resistance since the 1980s when it turned from a Western hippie travel destination into a mass and luxury tourism hub (Scheyvens, 2002; Telfer and Sharpley, 2008). Local resistance movements have ranged from moderate research-based environmental advocacy groups, such as the Goa Foundation, to more activist groups, including the Vigilant Goan's Army (*Jagrut Goenkaranchi Fouz* – JGF) (Routledge, 2001; Scheyvens, 2002). Saldanha (2002) traced back the particular assemblages of resistance to tourism in Goa to local society's specific identity shaped by four and a half centuries of Portuguese rule and the state's relative recent integration into the Indian nation state.

Construction work on the new greenfield international Mopa Airport in Goa began in 2017, following a decade-long land conflict centred on the state government's acquisition of around 800 hectares of land on which the airport will be constructed and the social and environmental implications of the airport project (Nielsen, 2017). In the 15 months from April 2017 to June 2018, the Forest Department granted permission to fell more than 21,000 trees at the Mopa air-port construction site (*The Economic Times*, 2018). This is remarkable because the Government of Goa's Environmental Impact Assessment (EIA) draft report of 2014 stated that "land in the project site is largely uncultivated with sparse vegetation" (Government of Goa, 2014, p. 37). Several photos in the EIA report depict the site as barren wasteland with only a few scattered shrubs.

Environmental groups and human rights advocates criticised that the EIA report – prepared by a public-sector consulting company – was an obvious attempt to downplay the environmental and social consequences of the project. In fact, the report claimed that "a very small percentage of the total working population in the North Goa district [the airport site] engaged in agriculture" (Government of Goa, 2014, p. 39). While the EIA report indirectly acknowl-edged that nearly 7,300 people in six villages had been affected by compulsory land acquisition, it emphasised that "these six villages form just around 1% of the total district's population" (ibid). At the time when the EIA report was published, all land had already been acquired by the government from previous landowners (Nielsen, 2017). The EIA report implied that the resettlement would actually bring "more benefits for the local people" through "employment opportunities for local skilled and unskilled people, development of infrastructure, commu-nications facility, drinking water supply, health" as well as unspecified "social and cultural development" (Government of Goa, 2014, p. xi).

The mandatory public hearing of the draft EIA report in February 2015 was attended by various groups opposing the airport construction and heated debates ensued which at some point required the intervention of the police force (Nielsen, 2017). Submissions critiqued the alleged manipulation of social and environmental data in the EIA draft report, questioned the plans of the authorities to use agricultural irrigation water as water supply for the air-port and referred to painful experiences with earlier resettlements from mining sites (*Bharat Mukti Morcha*, 2015). Hundreds of signed statements detailed their historical connection to the place and highlighted its ecological diversity, its importance for food production and other essential livelihood activities, and the cultural significance of various religious sites that would be destroyed by this greenfield development project (Nielsen, 2017).

The revised EIA report (Government of Goa, 2015) acknowledged that "forests [are] the predominant land use [in] the area" (p. iv) and contained a table with "environmentally sensitive areas" (p. 8) including reserved forests, wetlands, water bodies and archaeologically important places, but reiterated that the Mopa Airport would have "significant positive impact on employment and occupation" (p. x) and "that the airport development will not only increase and support tourism, but also accommodate the projected growth in business travel and cargo movements in Goa" (p. 106). Based on the revised EIA, the project obtained environmental clearance in October 2015 (Rajagopal, 2020).

This example shows how the Mopa Airport has been promoted by authorities and consultants as both a necessity and an economic boon for the state by emphasising how it serves "both the tourism and business markets and keeping pace with the growing air travel segment in India" (Government of Goa, 2015, p. ii). Instead of scrutinising the socio-ecological viability of the project and making sure that environmental and social safeguards would be adhered to, the EIA process rubber-stamped the project that had already been approved in principle by the government. This was underpinned by a gleaming rhetoric of 'greenfield development' for a project that actually will have a massive impact on the local environment and on dislocated communities. While the environmental clearance was temporarily suspended in 2019 due to processual flaws, the Supreme Court lifted the suspension in early 2020 after the concessionaire, GMR Goa International Airport Limited, committed to a zero-carbon programme for the airport's construction and operation by planting a total of 550,000 trees (Rajagopal, 2020). It is interesting to note that the ostensibly green credentials of the revised project design were given far more weight in the Supreme Court's decision than human rights aspects and social impacts.

Airport expansion and resettlement in Luang Prabang, Lao PDR: the problematic involvement of national and multilateral development banks

The landlocked Southeast Asian nation of Lao PDR is a relative newcomer to the international tourism industry. The socialist country has officially welcomed international tourists only since 1989 under its economic reform programme. By the mid-1990s, tourism had become a major priority for economic development and by 2005 it reached the one-million international visitor mark (Harrison and Schipani, 2007). With growing opportunities from tourism, a small elite has enriched itself through the construction of golf courses, casinos and integrated resort facilities (Sims and Winter, 2015).

As a UNESCO World Heritage Site, the town of Luang Prabang is Lao PDR's key northern node for increased regional connectivity and one of the country's most popular tourist destinations that has seen rapid investments in new hotels, restaurants, and other hospitality services. Yet it is also a city that is not well connected by road to other major cities in the Greater Mekong Subregion (GMS) and hence is heavily dependent on its small airport for most of its tourism and trade flows. In order to upgrade and expand the existing airport facilities, the Government of Laos requested a loan of US$57.8 million from the Export-Import Bank of China (cf. Box 9.2), while China CAMC Engineering Co Ltd was responsible for the survey, design and construction of the rebuilding project (Dreher et al., 2017). Meanwhile, the Asian Development Bank (ADB) – an international financial institution that is committed to enhancing regional economic integration in the GMS (cf. Box 9.2) – provided important discursive justification for the project through the provision of a technical assistance report (ADB, 2008; Sims, 2015).

Box 9.2 The Asian Development Bank and the Export-Import Bank of China as funding institutions for tourism infrastructure development

The ADB Is an International financial institution (IFI) that is committed to regional development and poverty alleviation. In one of its mission statements, ADB affirms its commitment to "achieving a prosperous, inclusive, resilient, and sustainable Asia and Pacific, while sustaining its efforts to eradicate extreme povert". Established in 1966, ADB has currently 68 shareholding members, of which 48 hail from the Asia-Pacific region. In 2017, ADB operations totalled 32.2 billion, including $11.9 billion in co-financing. One of ADB's foremost strategies for poverty alleviation is increased economic regionalism through regional cooperation and integration which are considered 'critical for Asia's march towards prosperity' (Sims, 2015). ADB regards tourism as a pro-poor strategy in Lao PDR and has provided substantial funds to the so-called Mekong Tourism Development Project in the early 2000s. A strong emphasis of the project has been on (1) building or improving roads and airports; (2) development of pro-poor, community-based tourism; and (3) improving cross-border tourism facilities in this landlocked country. In October 2018, ADB and the Lao government signed a US$48 million grant agreement for the Second Greater Mekong Subregion (GMS) Tourism Infrastructure Project, which aims at supporting road access, improving tourism-related services and enhancing sustainable tourism management.

The Export-Import Bank of China is a state-funded and state-owned policy bank with the status of an independent legal entity. It is a bank under the direct jurisdiction of the State Council and dedicated to supporting China's foreign trade, investment and international economic cooperation. In 2017, its foreign trade loans amounted to nearly US$145.7 billion, while its overseas investment loans totalled about US$38.1 billion. The bank's main emphasis overseas is on supporting large-scale infrastructure projects, such as a bridges, roads, airports, ports and dams. Yet, it has recently broadened its portfolio to strengthen its investment in the tourism industry. In Southeast Asia, one of the bank's investment groups has prepared to launch a fund of up to US$3 billion that will not only target infrastructure and telecoms but also travel businesses.

Source: Sims, 2015; Export-Import Bank of China, 2018; Harrison and Schipani, 2007

The Technical Assistance Report (TAR) anticipated an increase of 125 per cent of foreign tourists within the first three years of the completion of the airport upgrade and expected the expansion of its size to enhance employment, stimulate trade, promote tourism, and diversify livelihoods (Sims, 2015). Yet the TAR also conceded that the upgrade would directly affect a total of 814 households (132 of them headed by women) through loss of land and/or assets of which 424 households needed to be relocated to a new site (ADB, 2008). The airport upgrade would not only lead to the demolition of private homes, but also destroy businesses (including shops and brick factories), schools and other educational facilities, religious sites, agricultural plots and fruit trees (Sims, 2015). The costs for resettlement compensation and environmental improvements at the relocation site were estimated at approximately US$14.5 million (ADB, 2008). As the TAR recognised that the relocation carried the risk of undermining local livelihoods, the ADB developed a 'comprehensive land acquisition and resettlement action plan' that provided detailed guidance on how the resettlement process should be conducted (Sims, 2015). Provided that the action plan was to be implemented, the social and economic impact of the relocation was expected to be 'minor' and project-affected residents would be 'fully resettled and compensated' (ADB, 2008).

What was largely left out from this account were the following issues:

- the negotiated compensation funding provided by Export-Impact Bank of China was more than US$ 10 million less than the expected resettlement costs;
- the Lao Government – which implemented the relocation process – had a bad track record of earlier relocations and lacked interest in applying ADB's resettlement guidelines; and
- relocated agricultural households would face difficulties in re-establishing rice fields, gardens and fruit trees in the new location.

(Sims, 2015)

At the relocation site, almost all resettled families received a 180 m^2 land parcel, as opposed to the minimum of 400 m^2 outlined in the ADB's resettlement policy and regardless of how much land they had previously owned (Sims and Winter, 2015). Land was allocated on a lottery basis, and most families were relocated before they had the opportunity to build new homes and had to live temporarily in makeshift tents, some for more than a year (Sims, 2015). The provincial government that managed the resettlement process provided no compensation for fruit trees and other lost assets or for the income opportunities lost due to the need to rebuild homes at the new location (Sims and Winter, 2015). In contrast to the ADB resettlement plan which recognised that water and electricity should be in place prior to relocation, it took two months for these services to be supplied, and several households who had made an unlucky draw regarding their plot in the new site were still without piped water and electricity one year after the relocation (Sims, 2015).

Sims (2015) reported that resettled residents complained that their new plots of land were substantially smaller than their former landholdings and less suitable for agriculture and that their compensation payments were insufficient to rebuild their homes in the same quality as their former ones. Resettled people did not receive vocational training and food provisions – as recommended in the TAR – despite the villagers' urgent request for food aid in the first year after the resettlement (Sims and Winter, 2015). While the ADB cannot be held accountable for the poor implementation of the resettlement process, the example questions the pro-poor rhetoric of tourism infrastructure development that is common among international financial institutions and national devel opment banks. The enhanced mobility of foreign tourists and domestic elites contrasts sharply with the drastically reduced physical, social and economic mobility of those people that had to move out of tourism development's way.

Large-scale tourism infrastructure development: the contentious case of the Mayan Train megaproject in Southern Mexico

Tourism in the Yucatán Peninsula has been dominated by the four Ss 'sun, sand, sea and sex' from the inception of the Cancún megaresort in the mid-1970s (cf. Chapter 3) until the late 2010s. While beach tourism has spread further along the so-called 'Riviera Maya', much fewer tourists have visited the ecological, archaeological and cultural treasures in the interior of the peninsula. Upon his inauguration in late 2018, Mexico's new left-wing pre-sident Andrés Manuel López Obrador set out an ambitious agenda to open up the most remote areas of Yucatan to mass tourism. His signature megaproject, the 'Mayan Train' – scheduled for completion in 2023 – consists of a route of more than 1,500 kilometres connecting five of the country's southern states, namely Chiapas, Tabasco, Campeche, Quintana Roo and Yucatán (Camargo and Vázquez Maguirre, 2020). The planned train track will pass through sev-eral protected areas, including the UNESCO-designated Calakmul Biosphere Reserve, with its more than $7,000km^2$ area being the second-largest expanse of tropical forest in the Americas after the Amazon and providing habitats for iconic species such as jaguars, pumas and tapirs (Godoy, 2018; Camargo and Vázquez Maguirre, 2020). Since the Ancient Maya City and the Protected Forests of Calakmul were declared a Mixed World Heritage Site, tourism in the area has increased substantially; between 2012 to 2016, the number of tourists almost doubled (UNESCO, 2018). The 'Mayan Train' project is expected to increase tourist numbers exponentially, with a projected 8,000 tourists per day by 2023 (Beatley and Edwards, 2020).

Only a couple of weeks after his swearing-in ceremony, the President – flanked by local politicians, business people and Indigenous community lea-ders – participated in a Mayan ritual requesting the permission of *Pacha Mama* (Mother Earth) to build this infrastructure project that is projected to cost upward of US$6.5 billion (Pskowski, 2019). The location of the ritual – the city of Palenque adjacent to an important Mayan archaeological site – was carefully chosen. It is one of 15 stations along the track that are to be developed by a host of private investors. In addressing the participants, the president declared the project "an act of justice, because this region has been the most abandoned" (cited in Pskowski, 2019, n.p.). He claimed that the project would create 80,000 jobs in 2020 and 150,000 in 2021 (*Mexico News Daily*, 2020a).

The Yucatán Peninsula is predominantly inhabited by Mayans along with several other Indigenous groups, namely the Tseltal, Ch'on, Tsotsil, Zoque and Chontal (Camargo and Vázquez Maguirre, 2020). The planned project has created deep rifts among Indigenous peoples in the area. While many have hopes to be lifted out of poverty with the availability of jobs in the tourism sector and other industries that will benefit from better connectivity, several Indigenous associations are vehemently opposing the project, some calling it 'an act of war' (Beatley and Edwards, 2020) and 'a project of death' (*The Yucatan Times*, 2019). These groups accuse the Mexican government of infringing upon their Indigenous territories and communal lands (*ejidos*) and of insufficient consultation and information about the nature of the project. The fiercest opposition to the megaproject comes from the far-left Zapatista Army of National Liberation (*Ejército Zapatista de Liberación Nacional* – EZLN), whose members have vowed that they are willing to die as "protectors of the earth" in their fight against the megaproject (IWGIA, 2020, p. 442).

The Mexican Office of the United Nations High Commissioner for Human Rights attested that the project failed to meet international standards of consulting Indigenous people and seeking their free, prior and informed consent and that information meetings "referred only to the potential benefits of the project and not to the negative impacts it might cause" (UNHCHR Mexico Office, 2019, n.p.). The human rights body noted the very low turnout of voters – with only 2.8 per cent of the local Indigenous populations casting their vote – underrepresentation of women in consultation processes, and the strategic promise of health services and water supply to the region in exchange for support for the project (Beatley and Edwards, 2020). Indeed, the project appears to fall short of meeting many of the legal standards and principles set out in various binding national and international legal frameworks (cf. Box 9.3). No environmental, social and cultural impact evaluation has been conducted for the entirety of the project, and some of the few environmental impact assessments that have been completed for portions of the project have not been made public (Verza, 2020). Greenpeace México, the Mexican Centre for Environmental Law and several university academics allege that biodiversity in protected areas will suffer from severe deforestation and land degradation as a result of the construction and operation of the 'Maya Train' (*Mexico News Daily*, 2020a).

Box 9.3 National and international legal frameworks with relevance to the 'Mayan Train' megaproject

Constitution of Mexico (Constitución Política de los Estados Unidos Mexicanos)

Article 2 guarantees Indigenous peoples the right to self-determination and the autonomy to decide their internal forms of co-existence and social, economic, political, and cultural organisation.

National Institute of Indigenous Peoples' Law (Ley del Instituto Nacional de los Pueblos Indígenas)

Article 6 guarantees Indigenous peoples the right to consultation and free, prior and informed consent when Mexico's federal government promotes legal reforms and administrative acts that are likely to affect them.

United Nations Declaration on the Rights of Indigenous Peoples (UNDRIP)

Article 26 requires states to give legal recognition and protection to Indigenous peoples' lands, territories and resources which they have traditionally owned, occupied or otherwise used or acquired.

Article 32 supports Indigenous peoples' right to self-determination, consultation and free, prior and informed consent prior to approval of any project that may have an impact on them.

Convention 169 of the International Labor Organization (ILO)

Article 7 supports Indigenous peoples' rights "to decide their own priorities for the process of development as it affects their lives, beliefs, institutions and spiritual well-being and the lands they occupy or otherwise use, and to exercise control, to the extent possible, over their own economic, social and cultural development". The article further asks governments to "ensure that, whenever appropriate, studies are carried out, in co-operation with the peoples concerned, to assess the social, spiritual, cultural and environmental impact on them of planned development activities."

Article 13 requires governments to respect the cultural and spiritual value that Indigenous peoples attach to their lands and territories, as well as the collective nature of that relationship.

Source: Pskowski, 2019; Beatley and Edwards, 2020; Camargo and Vázquez Maguirre, 2020

The project's implementing body, the National Fund for Tourism Development (FONATUR) conceded that it had skipped some of the legally required steps and had accelerated the timeline, but argued that it will minimise the environmental and social impacts by mostly using existing train tracks and public land (Verza, 2020). Yet representatives of the Assembly of Defenders of the Mayan Territory and the Mexican Civil Council for Sustainable Forestry (CCMSS) maintain that the project has already caused intense land speculation and that government agencies are pressuring Indigenous communities to give up their customary land around the future train stations and urban growth centres (*The Yucatan Times*, 2019). A particularly perfidious strategy of FONATUR appears to be the promise of providing better water supplies to this notoriously water-scarce region on the condition that the megaproject can go ahead (UNHCHR Mexico Office, 2019; Beatley and Edwards, 2020).

Resistance to such megaprojects as the 'Maya Train' can be dangerous in Mexico, where violence against environmentalists and land rights activists is rampant and often goes unpunished (Beatley and Edwards, 2020). In 2019, 23 Mexican human rights activists and land defenders were killed according to Front Line Defenders (2020), a Dublin-based non-governmental organisation. One Mayan activist who openly expressed criticism of the project and the lack of Indigenous consultation reportedly received a death threat in December 2019 (Beatley and Edwards, 2020).

In May 2020, Mexico's Human Rights Commission called on the government to temporarily halt the construction of the 'Maya Train', with the argument that non-essential work on the project would risk the exposure of Indigenous communities to COVID-19 (*Yucatan Expat Life*, 2020). Due to similar concerns, a federal judge ordered a suspension of the construction work in June 2020 for as long as the COVID-19 epidemic remains a threat to Indigenous people's health (*Mexico News Daily*, 2020).

Concluding Remarks

Massive relocation of residents has been a persistent feature of sports mega-events and large-scale tourism infrastructure projects. Displacement was particularly pronounced in the run-up to the 2010 Men's FIFA World Cup in South Africa, the 2014 Men's FIFA World Cup in Brazil and the Rio 2016 Summer Olympic Games. All three major sports events had in common that disadvantaged groups – urban squatters, homeless people and street vendors – were further marginalised and disenfranchised by the construction of stadiums and urban renewal schemes. Legitimised by a 'public purpose' and 'tourism and security' discourse, the economic and social conditions of the urban poor deteriorated further in the process, while urban elites and large corporations benefitted the most, as large funds were channelled into the production of new urban spaces for the privileged. Protests and emerging resistance alliances against the adverse impacts of these sports mega-events were dismissed by government officials as unpatriotic and treasonous. As the examples from India, Laos and Mexico have shown, large-scale tourism infrastructure projects for airports and railways have also been associated with involuntary relocation, forced land acquisition, destruction of livelihoods and erasure of local cultures. The case of resettlement for the airport upgrade in Luang Prabang testifies to the discrepancy between relocation planning and practice, whereby resettled minority populations find themselves in much worse living conditions than before, despite the legitimising rhetoric of state agencies and international financial institutions that underpins these developments. Resistance to such large-scale tourism infrastructure projects is almost non-existent under the authoritarian regime of the Government of Laos. Yet, even in democracies, such as India and Mexico, where resistance movements have a long tradition, protests have had very limited success rates.

The next chapter will synthesise the findings from Chapters 3–9. It will draw on Harvey's (2006) mechanisms involved in processes of 'accumulation by dispossession', privatisation, financialisation, management and manipulation of crises, and state redistributions. Then it will explore four practices of dispossession, namely eviction, enclosure, extraction and erasure, that have been a work in the 31 case studies discussed. The chapter will conclude with an examination of resistance movements against tourism-related land grabs and displacement.

References

ADB (2008) *Lao People's Democratic Republic: Greater Mekong Subregion Louangphrabang Airport Improvement Project*. Asian Development Bank (ADB): Manila.

Baka, J. (2013) 'The political construction of wasteland', *Development and Change*, 44 (2): 409–428.

Bennike, R. (2017) 'Frontier commodification: Governing land, labour and leisure in Darjeeling, India', *South Asia: Journal of South Asian Studies*, 40 (2): 256–271.

Camargo, B.A. and Vázquez-Maguirre, M. (2020) 'Humanism, dignity and indigenous justice: The Mayan Train megaproject, Mexico', *Journal of Sustainable Tourism*. http s://doi.org/10.1080/09669582.2020.175870.

COHRE (2007) *Fair Play for Housing Rights: Mega-Events, Olympic Games and Housing Rights*. Available at: www.ruig-gian.org/ressources/Report%20Fair%20Play %20FINAL%20FINAL%20070531.pdf (accessed 14 February 2019).

COHRE (2009) *N2 Gateway Project: Housing Rights Violations as "Development" in South Africa*. Available at: https://westerncapeantieviction.files.wordpress.com/2009/09/ 090911-n2-gateway-project-report.pdf (accessed 14 February 2019).

Comitê Popular da Copa e das Olimpíadas do Rio de Janeiro (2014) *Megaeventos e Violações dos Direitos Humanos no Rio de Janeiro* [Megaevents and Human Rights Violations in Rio de Janeiro]. Comitê Popular da Copa e Olimpíadas do Rio de Janeiro: Rio de Janeiro. Available at: https://comitepopulario.files.wordpress.com/ 2014/06/dossiecomiterio2014_web.pdf (accessed 14 February 2019).

Comitê Popular da Copa e das Olimpíadas do Rio de Janeiro (2015) *Olimpíada Rio 2016: os jogos da exclusão*. [The 2016 Rio Olympics: The Exclusion Games]. Comitê Popular da Copa e Olimpíadas do Rio de Janeiro: Rio de Janeiro.

Cornelissen, S. (2011) Mega event securitisation in a third world setting: Glocal processes and ramifications during the 2010 FIFA World Cup. *Urban Studies*, 48 (15), 3221–3240.

Cummings, J. (2015) 'Confronting favela chic: The gentrification of informal settlements in Rio de Janeiro, Brazil' in Lees, L., Shin, H. and Lopez-Morales, E. (eds) *Global Gentrifications: Uneven Development and Displacement*. Policy Press: Bristol, pp. 81–99.

Davis, L.K. (2011) 'International events and mass evictions: A longer view', *International Journal of Urban and Regional Research*, 35: 582–599.

Dreher, A., Fuchs, A., Parks, B.C., Strange, A.M. and Tierney, M.J. (2017) *Aid, China, and Growth: Evidence from a New Global Development Finance Dataset*. AidData Working Paper #46. AidData: Williamsburg, VA.

Export-Import Bank of China (2018) *Annual Report 2017.* Available at: http://english. eximbank.gov.cn/upload/accessory/201812/201812415543010122.pdf (accessed 14 January 2019).

Front Line Defenders (2020) *Front Line Defenders Global Analysis 2019.* Available at: www.frontlinedefenders.org/en/resource-publication/global-analysis-2019 (accessed 15 July 2020).

Gaffney, C. (2010) 'Mega-events and socio-spatial dynamics in Rio de Janeiro, 1919–2016', *Journal of Latin American Geography,* 9 (1), 7–29.

Gaffney, C. (2016) 'Gentrifications in pre-Olympic Rio de Janeiro', *Urban Geography,* 37: 1132–1153.

Gerrard, P. (2004) *Integrating Wetland Ecosystem Values into Urban Planning: The Case of That Luang Marsh, Vientiane, Lao PDR .* IUCN – The World Conservation Union and WWF – Lao Office: Vientiane.

Gobierno de la Ciudad Autónoma de Buenos Aires [Autonomous City Government of Buenos Aires] (2017) *Transformación del Distrito* [Transformation of the District]. Available at: www.buenosaires.gob.ar/distritodelasartes/transformacion-deldistrito (accessed 16 February 2019).

Government of Goa (2014) *Environmental Impact Assessment Study for the Proposed Greenfield Airport Construction at Mopa, Goa – Draft Version for Consultation.* Government of Goa: India.

Government of Goa (2015) *Environmental Impact Assessment Study for the Proposed Greenfield Airport Construction at Mopa, Goa – Final Version.* Government of Goa: India.

Government of India (2014) 'Amendments made in the right to fair compensation and transparency in Land Acquisition, Rehabilitation and Resettlement Act 2013', Press Release, 29 December. Available at: http://pib.nic.in/newsite/PrintRelease.aspx? relid=114190 (accessed 23 April 2015).

Harrison, D. and Schipani, S. (2007) 'Lao tourism and poverty alleviation: Community-based tourism and the private sector', *Current Issues in Tourism,* 10 (2–3): 194–230.

Harvey, D. (2006) 'Neo-liberalism as creative destruction', *Geografiska Annaler,* 88B (2): 145–158.

IWGIA (2020) *The Indigenous World 2020.* International Work Group on Indigenous Affairs (IWGIA): Copenhagen.

Janoschka, M. and Sequera, J. (2016) 'Gentrification in Latin America: Addressing the politics and geographies of displacement', *Urban Geography,* 37 (8): 1175–1194.

Lenskyj, H.J. (2002) *The Best Olympics Ever? Social Impacts of Sydney 2000.* State University of New York Press: Albany.

Maharaj, B. (2015) 'The turn of the south? Social and economic impacts of mega-events in India, Brazil and South Africa', *Local Economy,* 30 (8): 983–999.

McGuirk, J. (2016) 'Failing the informal city: How Rio de Janeiro's mega sporting events derailed the legacy of Favela-Bairro', *Architectural Design,* 86 (3): 40–47.

Mteki, N., Murayama, T. and Nishikizawa, S. (2017) 'Social impacts induced by a development project in Tanzania: A case of airport expansion', *Impact Assessment and Project Appraisal,* 35 (4): 272–283.

Müller, M. (2015) 'The Mega-Event Syndrome: Why so much goes wrong in mega-event planning and what to do about it', *Journal of the American Planning Association,* 81 (1): 6–17.

Müller, M. and Gaffney, C. (2018) 'Comparing the urban impacts of the FIFA World Cup and Olympic Games from 2010 to 2016', *Journal of Sport and Social Issues,* 42 (4): 247–269.

Neef, A. (2014) 'Law and development implications of transnational land acquisitions: Introduction', *Law and Development Review*, 7 (2): 187–205.

Neef, A. and Singer, J. (2015) 'Development-induced displacement in Asia: conflicts, risks and resilience', *Development in Practice*, 25 (5): 601–611.

Newton, C. (2009) 'The reverse side of the medal: About the 2010 FIFA World Cup and the beautification of the N2 in Cape Town', *Urban Forum*, 20: 93–108.

Nielsen, K.B. (2017) 'Unclean slates: greenfield development, land dispossession and EIA struggles in Goa', *South Asia: Journal of South Asian Studies*, 40 (4): 844–861.

Porter, L. (2009) 'Planning displacement: The real legacy of major sporting events', *Planning Theory & Practice*, 10 (3): 395–399.

Raco, M. and Tunney, E. (2010) 'Visibilities and invisibilities in urban development: Small business communities and the London Olympics 2012', *Urban Studies*, 47: 2069–2091.

Routledge, P. (2001) '"Selling the rain", resisting the sale: Resistant identities and the conflict over tourism in Goa', *Social & Cultural Geography*, 2 (2): 221–240.

Saldanha, A. (2002) 'Identity, spatiality and post-colonial resistance: Geographies of the tourism critique in Goa', *Current Issues in Tourism*, 5 (2): 94–111.

Scheyvens, R. (2002) *Tourism for Development: Empowering Communities*. Pearson Education Limited: London.

Silvestre, G. and de Oliveira, N.G. (2012) 'The revanchist logic of mega-events: community displacement in Rio de Janeiro's West End', *Visual Studies*, 27 (2): 204–210.

Sims, K. (2015) 'The Asian Development Bank and the production of poverty: Neoliberalism, technocratic modernization and land dispossession in the Greater Mekong Subregion', *Singapore Journal of Tropical Geography*, 36: 112–126.

Sims, K. and Winter, T. (2015) 'In the slipstream of development: World Heritage and development-induced displacement in Laos' in Labadi, S. and Logan, W. (eds) *Urban Heritage, Development and Sustainability: International Frameworks, National and Local Governance*. Routledge: London, New York, pp. 23–36.

Steinbrink, M., Haferburg, C. and Ley, A. (2011) 'Festivalisation and urban renewal in the global South: Socio-spatial consequences of the 2010 FIFA World Cup', *South African Geographical Journal*, 93 (1): 15–28.

Stuart-Fox, M. (2009) 'Laos: The Chinese connection', *Southeast Asian Affairs*, 2009: 141–169.

Telfer, D.J. and Sharpley, R. (2008) *Tourism and Development in the Developing World*. Routledge: London, New York.

Warrall, S. (2016) 'There's a dark history behind the glittering Olympic Games', *National Geographic*. Available at: https://nationalgeographic.com/2016/07/olympic-games-history-rio-david-goldblatt/ (accessed 17 February 2019).

Watt, P. (2013) '"It's not for us": Regeneration, the 2012 Olympics and the gentrification of East London', *City*, 17: 99–118.

Media sources/websites

Beatley, M. and Edwards, S. (2020) 'Is Mexico's "Mayan Train" a boondoggle?', *The Nation*, 22 May. Available at: www.thenation.com/article/world/is-mexicos-mayan-train-a-boondoggle/ (accessed 20 July 2020).

Bharat Mukti Morcha (2015) 'Objections to Mopa Greenfield Airport massive land grab', 28 January. Available at: http://bharatmukti.blogspot.com/2015/01/objections-to-mopa-greenfield-airport.html (accessed 7 January 2019).

Boliek, B. (2016) 'Lao Prime Minister moves to stop land-for-capital deals', *Radio Free Asia*, 16 December. Available at: www.rfa.org/english/news/laos/lao-prime-minister-moves-to-stop-12162016123757.html (accessed 10 September 2018).

Dalei, N.N. and Singh, D.P. (2015) 'Role of greenfield airports in greening the India economy', *Foreign Policy News*, 1 December. Available at: http://foreignpolicynews.org/2015/12/01/role-of-greenfield-airports-in-greening-the-india-economy/ (accessed 7 January 2019).

Economist Intelligence Unit (2012) 'Land concessions come under scrutiny', 1 June. Available at: http://country.eiu.com/article.aspx?articleid=1289097113&Country=Laos&topic=Politics&subtopic=Rec_1 (accessed 22 August 2020).

Finney, R. (2020) 'Lao capital residents fight land grab, reject offered compensation', *Radio Free Asia*, 16 July. Available at: www.southeastasianews.net/news/265798618/lao-capital-residents-fight-land-grab-reject-offered-compensation (accessed 22 August 2020).

Firstpost (2014) 'India planning 10–15 greenfield airports in non-metros. 20 December. Available at: www.firstpost.com/business/india-planning-10-15-greenfield-airports-in-non-metros-482964.html (accessed 12 January 2019).

Fuller, T. (2009) 'Laos stumbles on path to sporting glory', *New York Times*, 5 October. Available at: www.nytimes.com/2009/10/06/world/asia/06laos.html (accessed 13 January 2019).

Gerin, R. (2019) 'Lao capital villagers balk at below-market payment for land lost to expressway', *Radio Free Asia*, 14 August. Available at: www.rfa.org/english/news/laos/lao-capital-villagers-balk-08142019151459.html (accessed 22 August 2020).

Godoy, E. (2018) 'Local communities in Mexico question benefits of Mayan Train', *Inter Press Service News Agency*, 17 December. Available at www.ipsnews.net/2018/12/local-communities-question-benefits-mayan-train-southern-mexico/ (accessed 20 July 2020).

IATA (2016) 'IATA forecasts passenger demand to double over 20 years', International Air Transport Association (IATA), Press Release, 18 October. Available at: www.iata.org/pressroom/pr/Pages/2016-10-18-02.aspx (accessed 12 January 2019).

Kuronuma, Y. (2018) 'India plans to build 100 more airports for 1bn flyers by 2035', *Nikkei Asian Review*, 12 April. Available at: https://asia.nikkei.com/Business/Business-Trends/India-plans-to-build-100-more-airports-for-1bn-flyers-by-2035 (accessed 12 January 2019).

Mexico News Daily (2020a) 'Indigenous groups and others launch new broadside against Maya Train', 3 June. Available at: https://mexiconewsdaily.com/news/new-broadside-against-maya-train/ (accessed 20 July 2020).

Mexico News Daily (2020b) 'Judge orders definitive suspension of train construction due to virus', 24 June. Available at: https://mexiconewsdaily.com/news/coronavirus/judge-orders-definitive-suspension-of-train-construction-due-to-virus/ (accessed 20 July 2020).

Phillips, T. (2005) 'Blood, sweat and fears in favelas of Rio', *The Guardian*, 29 October. Available at: www.theguardian.com/world/2005/oct/29/brazil.mainsection (accessed 23 August 2020).

Pskowski, M. (2019) 'Mexico's "Mayan Train" is bound for controversy', *Bloomberg*, 23 February. Available at: www.bloomberg.com/news/articles/2019-02-22/mexico-s-yucatan-train-brings-promise-of-a-tourism-boom (accessed 20 July 2020).

Rajagopal, K. (2020) 'SC lifts EC suspension on Mopa Airport', *The Hindu*, 17 January. Available at: www.thehindu.com/news/national/other-states/sc-lifts-ec-suspension-on-goa-mopa-airport/article30580464.ece (accessed 22 August 2020).

RAN (n.d.) 'Japan's impact on forests: 2020 Olympics', Available at: https://act.ran. org/japan (accessed 12 January 2019).

Schuettler, D. (2008) 'China land deal rankles Laos capital', *Reuters*, 7 April. Available at: www.reuters.com/article/us-laos-chinatown/china-land-deal-rankles-laos-capita l-idUSBKK24347820080407?feedType=RSS&feedName=inDepthNews (accessed 10 January 2019).

Sen, S. (2015) 'Govt projects Rs24,000-cr cost to develop 14 greenfield airports', *The Financial Express*, 11 September. Available at: www.financialexpress.com/economy/ govt-projects-rs-24000-cr-cost-to-develop-14-greenfield-airports/133915/ (accessed 7 January 2019).

Talbot, A. (2018) 'Two years after Rio Olympics, Vila Autódromo celebrates "glorious victory," looks toward the future', 11 August. Available at: www.rioonwatch.org/? p=46080 (accessed 17 February 2019).

The Economic Times (2018) 'Goa's environment faces problems from infrastructure boom', 13 September. Available at: https://economictimes.indiatimes.com/topic/ Mopa-Greenfield-airport (accessed 12 January 2019).

The Yucatan Times (2019) 'The Maya Train a "project of death" – indigenous representa- tives claim', 5 July. Available at: www.theyucatantimes.com/2019/07/the-mayan-tra in-a-project-of-death-indigenous-representatives-claim/ (accessed 20 July 2020).

Tourism News Live (2018) 'China to build 216 new airports by 2035', *Tourism News Live*, 18 December. Available at: www.tourismnewslive.com/2018/12/18/china -to-build-216-new-airports-by-2035/ (accessed 12 January 2019).

UNHCHR Mexico Office (2019) 'El proceso de consulta indígena sobre el Tren Maya no ha cumplido con todos los estándares internacionales de derechos humanos en la materia: ONU-DH [The indigenous consultation process on the Maya Train has not met all international human rights standards in the field: UNHCHR]'. Available at: www.onu.org.mx/el-proceso-de-consulta-indigena-sobre-el-tren-maya-no-ha-cump lido-con-todos-los-estandares-internacionales-de-derechos-humanos-en-la-materia-o nu-dh/ (accessed 20 July 2020).

Vandenbrink, R. (2013) 'That Luang Marsh resident refuse to move', *Radio Free Asia*, 22 February. Available at: www.rfa.org/english/news/laos/that-luang-02222013173 942.html (accessed 22 August 2020).

Verza, M. (2020) 'Mexican President goes full-steam ahead with Mayan train', *Asso- ciated Press*, 4 June. Available at: https://abcnews.go.com/International/wireStory/m exicos-president-full-steam-ahead-mayan-train-71036966 (accessed 20 July 2020).

Yucatan Expat Life (2020) 'Mexican rights commission calls for halt to Mayan Train', 16 May. Available at: https://yucatanexpatlife.com/mexican-rights-commission-ca lls-for-halt-to-mayan-train/ (accessed 20 July 2020).

10 Tourism-related land grabs

Mechanisms, practices, impacts and resistance

This chapter attempts to synthesise the findings of the 31 case studies from 26 countries by looking at various types of mechanisms and practices involved in tourism-related land grabbing, examining the impacts on affected communities through a comparative perspective and distilling distinct forms of protest and resistance. The first section of this chapter draws on Harvey's (2006) enabling mechanisms of 'accumulation by dispossession'. The second part of the chapter examines four distinct practices of dispossession (introduced in Chapter 1) – eviction, enclosure, extraction, erasure – and illustrates them with summaries of selected case studies. Finally, the chapter explores community-level resistance to tourism-related land grabbing and displacement and introduces the concepts of 'assemblages of resistance' and 'resistance identities' which are major success factors of resistance movements against predatory tourism development.

Mechanisms

Harvey (2006) distinguishes four major mechanisms through which the global advance of neoliberalism has enabled 'accumulation by dispossession', i.e. the illegal, illegitimate and/or unethical appropriation of land and other resources in the hands of elites at the expense of the majority of the population. These four mechanisms are (1) privatisation, (2) financialisation, (3) management and manipulation of crises, and (4) state redistributions. In the following, the expression of these four mechanisms in the tourism sector is outlined with reference to the case studies presented in Chapters 3–9.

Privatisation

The privatisation and commodification of land and natural resources is a crucial step towards opening the door for land alienation. Communally owned land and natural resources – generally referred to as 'the commons' – are often seen as an impediment for investment and development. Most tourism businesses rely on secure, exclusive and alienable land rights that they can hold as private property and use as collateral for obtaining loans from the financial sector. Many governments in the Global South have followed advice from international legal experts to privatise both public and communally held land in order to provide incentives for domestic, foreign and multinational corporations to invest in land for tourism development. This has often been accompanied by the rolling back of restrictions on land sales and of regulations with regard to environmental and social standards. Through privatisation of land that was previously held as public land or under communal management, the state actively transfers "assets from the public and popular realms to the private and class-privileged domains" (Harvey, 2006, p. 153).

Proponents of privatisation argue that only land that is held as private property can be sold in a regular land market under a 'willing buyer, willing seller' arrangement (cf. de Soto, 2000). Yet they tend to overlook that many land sales in tourism zones are induced by excessively high land prices and/or are more accurately described as 'distress sales' whereby land owners risk losing the basis of their entire livelihood (cf. Chapter 3 for the case of the Philippines and Chapter 4 for the cases of Bali and Mauritius). In some cases, the process of privatisation predates the tourism boom (cf. the case of Costa Rica in Chapter 4), while in other cases privatisation has occurred as a direct impact of the tourism expansion (e.g. in the case of Honduras's North Coast, discussed in Chapter 5). A peculiar case is Vanuatu, a small island developing state in the Southwest Pacific, where private, long-term leases held by expatriate tourism investors have a much higher value as collateral than customary ownership of the Indigenous Ni-Vanuatu people (cf. Chapter 4). Finally, private ownership documented by land titles is not always a hedge against forced acquisitions and evictions, as exemplified by the case of Tulum on Mexico's Riviera Maya in Chapter 3.

Financialisation

The tourism boom and its adverse impacts on the land rights of local communities in many countries of the Global South is fed with large amounts of capital from commercial banks (cf. the case of post-disaster Honduras in Chapter 5) and national development banks (cf. the early years of Cancún's megaresort development and Cambodia's economic land concession in Chapter 3, and Lao PDR's airport expansion project Luang Prabang in Chapter 9). As Gibson (2019) states, excess capital of institutional and corporate investors continuously seeks new tourism spaces in which to reinvest. International financial institutions (IFIs) also play a major role in either directly or indirectly fuelling dispossession and evictions for large tourism projects (e.g. the funding of a regional connectivity project in the Chittagong Hill Tracts by the World Bank which will bring more visitors to tourist areas controlled by the Bangladesh military, described in Chapter 6). Financial support from IFIs for conservation and tourism projects have also fuelled resettlement from tourism zones and wildlife reserves, as exemplified by Germany's national development bank's (KfW) involvement in establishing the Limpopo National Park in

Mozambique as part of the Great Limpopo Transfrontier Park in southern Africa (Chapter 7).

In many places, the combined involvement of the global financial sector and local commercial banks in the provision of tourism infrastructure and building of megaresorts, has led to speculative land purchases and rapidly rising land prices that crowd out the local population from the land market. This is evident from residential tourism development in Mauritius and Costa Rica (Chapter 4) and the planned 'Maya Train' mega-project on the Yucatán peninsula, discussed in Chapter 9. In the northwestern coastal provinces of Costa Rica (Chapter 4) and along the Riviera Maya in southern Mexico (Chapter 3), affluent North American investors provide most of the financial capital that has driven the boom of resort and residential tourism.

Management and manipulation of crises

Many tourism booms in the Global South have been precipitated by some form of crisis, and this has not always occurred by accident. Through creating, managing or manipulating crises, governments in alliance with the corporate tourism sector can easily construct a rationale for tourism development as a pathway out of the crisis. This ties in with the various crisis discourses that have been deployed by many governments in the Global South as described in Chapter 2.

The government of Costa Rica considered the tourism sector as a new growth strategy when the agricultural sector was in an economic crisis (Chapter 4). Bali was promoted as a low-cost tourist destination following the 1997 Asian financial crisis (Benge and Neef, 2018; Chapter 4). Timor-Leste's government views tourism development as a strategy to avert budget deficits when the country is running out of its offshore oil and natural gas reserves. The Sri Lankan government employed tourism as a strategy to recover from both the 2004 Indian Ocean Tsunami and the country's long civil war (Chapter 6). In Honduras, tourism was first promoted after the 1998 Hurricane Mitch and then a decade later as part of a post-coup economic recovery process (Chapter 5). In all these cases, the sudden prioritisation of the tourism sector as a crucial – if not the only – crisis management strategy has enabled governments to enact new, investor-friendly legislation and attract large flows of foreign capital into their countries' tourism industry, with little regard for the land and resource rights of local communities living in areas with high tourism potential.

State redistributions

In order to boost tourism, governments employ a range of redistributive measures, often using taxpayers' money to subsidise tourism development through the provision of key infrastructure (roads, airports, ports, public water supply), as exemplified by the case of Labuan Bajo in Eastern Indonesia

(Chapter 4) and Goa in India (Chapter 9), where – apart from the construction of airports and other transportation infrastructure that have led to displacements – water supplies have also been diverted from local users (farmers, urban residents) to the growing tourism industry. In other cases, governments forfeit tax revenues by providing generous tax breaks for domestic and foreign tourism businesses, as evident from the cases presented from Mexico (Chapter 3), Costa Rica and Mauritius (Chapter 4) and Honduras (Chapter 5).

Some states in the Global South are also directly redistributing thousands of hectares of 'public' land to foreign corporations, e.g. in the form of economic land concessions, as exemplified by the case of Cambodia (Chapter 3). Governments may also redistribute control over land to state entities that traditionally have not been involved in the tourism industry, as evidenced by the case of military-driven tourism development and displacements in Sri Lanka and Bangladesh (Chapter 6).

In some cases, 'appropriation by dispossession' for tourism purposes combines several mechanisms. In the post-disaster case of Sicogon Island in the Philippines, discussed in Chapter 5, the government shifted the responsibility for recovery to a large corporation that used this form of 'state redistribution' to manipulate the crisis for its own economic goals of turning the island into a prime tourism destination at the expense of the disaster victims.

Practices and impacts of tourism-related land grabs

Drawing on selected case studies that are described in more detail in Chapters 3–9, this section looks at the four distinct *practices of dispossession* (cf. Devine and Ojeda, 2017; Neef et al., 2018) introduced in Chapter 1 and illustrates them with short summaries in a tabular form to show typical patterns.

Eviction

Land grabs for tourism purposes often lead to *forced evictions* which are defined as the "permanent or temporary removal against their will of individuals, families and/or communities from the homes and/or land which they occupy, without the provision of, and access to, appropriate forms of legal or other protection" (OHCHR, 1997, pp. 1–2). According to the special rapporteur on adequate housing forced evictions "constitute gross violations of a range of internationally recognized human rights, including the human rights to adequate housing, food, water, health, education, work, security of the person, security of the home, freedom from cruel, inhuman and degrading treatment, and freedom of movement" (OHCHR, 2018). In most cases, forced evictions violate various civil and political rights, such as the right to security, the right of freedom of movement and the right to peaceful enjoyment of possessions. Table 10.1 provides a compilation of typical examples from the case studies.

While not all evictions come with the use of direct force, there are other forms of physical displacement, such as involuntary displacement, resettlement

Table 10.1 Practices of dispossession in tourism – Eviction

Case study	Characteristics
Cambodia (Southeast Asia) Chapter 3	A large-scale tourism project by a Chinese corporation in Koh Kong Province forced hundreds of families from coastal land they had occupied for many decades; families were given no choice and only meagre compensation; resisting groups have faced violence by private security guards and the Cambodian military; families were resettled into the interior of the Botum Sakor National Park where illegal logging remains one of few options to sustain their livelihoods.
Bangladesh (South Asia) Chapter 6	Following a 20-year civil conflict in the Chittagong Hill Tracts, the military has maintained a strong presence to 'securitise' the region for tourism but also controlling a large part of the tourism sector itself. Several hundred indigenous families from the Jumma, Mro and Marma ethnic groups have been forcefully evicted from their land to make way for military-owned tourist resorts popular with domestic tourists.
Tanzania (East Africa) Chapter 7	The expansion of national parks and wildlife sanctuaries for safari tourism and trophy hunting has forced thousands of pastoralist Maasai from their customary lands; military and police have used brute force for evictions; no compensation and no alternative settlement areas have been provided and there are no institutionalised grievance mechanisms in place.
Colombia (South America) Chapter 7	Peasant settlers (*colones*) and fisherfolks in the Tayrona National Park were dispossessed and driven from their land by paramilitary 'clean-up' operations during the civil war to 'secure the area' for ecotourism development. A fishing community was evicted by park officials who destroyed the fishers' huts that have existed for half a century, while sparing luxurious private beach homes of local elites.
Peru (South America) Chapter 8	In the World Heritage 'City of Cuzco', the former capital of the Inca Empire, municipal authorities cleared the inner city of 'urban undesirables', such as street vendors and the urban poor, to provide foreign tourists with a sense of comfort and security. Thousands of street vendors were forcibly removed in the early 2000s. This was complemented by market-led gentrification, with investors buying up prime property in the city's centre.
Lao PDR (Southeast Asia) Chapter 9	The construction of the international airport in Luang Prabang, a famous World Heritage Site in the northern part of the country, triggered the involuntary displacement of several hundred ethnic minority families who received inadequate compensation, obtained smaller plots of land than they were originally promised and had to live without piped water and electricity for an extended period of time.
Brazil (South America) Chapter 9	Two consecutive sport mega-events in Brazil (the 2014 FIFA World Cup and the 2016 Rio Summer Olympics) were used by the state government of Rio de Janeiro to legitimise the eviction of thousands of informal residents from the city's *fávelas*. It also deployed Police Pacification Units (UPPs) – co-financed by the private sector – to 'pacify' unruly neighbourhoods and regain administrative control over urban spaces.

under 'induced volition', or 'voluntary' resettlement based on deceit, false hopes or lack of alternative options. These may not always be categories in a legal sense, but the distinction between those concepts involves a number of ethical questions, such as who defines whether a person or community leaves a certain place out of their free will and what kind of information is available prior to the displacement. In many cases, the immediate impact of physical displacement is a near-total loss of livelihood opportunities, even when compensation may provide temporary support for the transition to other livelihood options.

Enclosure

Even when people are not physically displaced by a land grab and are allowed to stay on the land they occupy, they may face various forms of *enclosure* of land and resources that they could hitherto use freely. Table 10.2 shows some typical examples from the case studies.

As Saarinen and Wall-Reinius (2019, p. 746) remind us, tourism is a form of land use, and gated communities and enclave-style resorts in particular are "manifestations of territorialization and privatization" and "create socio-spatial patterns of inclusions and exclusions". Thereby, tourism enclaves can have dramatic impacts on local people's access to essential resources, such as clean water (Gössling et al, 2012; Cole et al., 2020).

The examples discussed in previous chapters and summarised in Table 10.2 show that tourism-related enclosure may restrict people's access to a range of other natural resources, such as the foreshore, inshore fishing grounds or seaweed plantations, mangrove areas, communal pastures, and forests. Enclosures may also cut off people's access to vital local infrastructure, such as roads, electricity, schools and markets. New land use regulations associated with the delineation of a tourism zone or a conservation area may also constitute a form of enclosure. In sum, enclosure of resources previously held in common by social groups have an adverse impact on people's livelihoods and can constitute a serious form of economic, social and cultural displacement.

Extraction

Despite its 'feel-good' image and an increasingly vivid international discourse around 'sustainable tourism', the tourism sector is in many ways an extractive industry. Several international studies have recently pointed to the enormous water footprint of tourism (e.g. Gössling et al., 2012; Becken, 2014). Tourism consumes high amounts of water both during the construction process and the operation of facilities and services, including swimming pools, golf courses, laundry, spas, gardens and catering. A study in the water-stressed island of Zanzibar, Tanzania, found that hotels' average daily water use per room was close to 1,500 litres, 16 times higher than the daily water consumption of local

Table 10.2 Practices of dispossession in tourism – Enclosure

Case study	Characteristics
Philippines (Southeast Asia) Chapter 3	On Boracay Island (Western Visayas region) and in Hacienda Looc (Batangas province), the delineation of tourism economic zones and expansion of large-scale resort complexes has pushed Indigenous and non-indigenous communities to the fringes and rendered coastal areas and fishing grounds inaccessible to subsistence farmers and fisherfolks.
Vanuatu (South Pacific) Chapter 4	In the South Pacific island Efate, part of the archipelago of Vanuatu, rampant leases of customary-owned coastal areas to expatriate tourism developers has restricted local residents' access to beachfront areas and made near-shore fishing – previously important for women's livelihoods – difficult or even impossible.
Mauritius (Africa) Chapter 4	In the Black River District of this small island state, rapid expansion of the residential tourism market has led to the enclosure of beach areas by large-scale tourism enclaves and gated communities that exclude local residents from access to affordable housing and evokes sentiments of foreign domination, reminiscent of the colonial era.
Haiti (Caribbean) Chapter 5	An enclave resort owned by cruise company Royal Caribbean on the northern shores of the country bars locals from accessing beaches and inshore fisheries and allows only selected local vendors to sell souvenirs. On the southern island of Île à Vâche, airport and road constructions for a failed tourism project have enclosed smallholders' farmland, coconut plantations and communally held natural resources.
India (South Asia) Chapter 7	In the Similipal Tiger Reserve in the State of Odisha, communities that remained in the core zone of this wildlife sanctuary faced severe restrictions regarding their access to forest resources, particularly non-timber forest products (NTFPs); the ongoing enclosure of communally used resources by the park authorities is expected to 'motivate' tribal and indigenous peoples to leave the tiger reserve.
Guatemala (Central America) Chapter 8	The controversial Maya Security and Conservation Partnership program in the Maya Biosphere Reserve in northern Guatemala – promoted by an American archaeologist, a group of US senators and a Guatemalan corporate foundation – threatens to absorb portions of community-based forest concessions into national parks and crowd out local tourism initiatives through massive tourism infrastructure.

households (Tourism Concern, 2012). As the examples in Table 10.3 show, other resources that are often extracted to make way for tourism infrastructure or to build mega-resorts are fruit trees, mangrove forests and timber trees that play a vital role in sustaining local livelihoods.

As exemplified by the case of Cancún in Mexico's Yucatán Peninsula (cf. Table 10.3 and Chapter 3), sand mining has become a novel but increasingly common extractive practice in tourism. In the Maldives, for example, luxury resorts on private islands have reportedly extracted sand from adjacent islands to beautify, stabilise and protect their beaches for tourists, while exposing islands inhabited by the local population to the risk of storm surges and rising sea levels (pers. comm., Inaz Ahmad).

Table 10.3 Practices of dispossession in tourism – Extraction

Case study	Characteristics
Mexico (North America) Chapter 3	Around the Cancún megaresort and along the Riviera Maya on Mexico's Yucatán Peninsula, hundreds of hectares of mangrove forests have been removed to make way for beach resorts and other tourism developments. Sand has been mined from nearby islands and from the ocean floor to temporarily restore beaches after major storm events. Scarce freshwater resources are prioritised for the hotel industry and foreign tourists.
Timor-Leste (West Pacific) Chapter 3	Large-scale infrastructure projects for tourism development in the Oecusse-Ambeno enclave of Timor-Leste have removed fruit trees, gardens and wells from the indigenous Meti people without proper compensation. In addition to the economic loss of timber and fruit from extracted trees, residents also lost important markers of their ownership claims.
Indonesia (Southeast Asia) Chapter 4	On the island of Bali, Indonesia's prime tourist destination, and in emerging tourist centres in East Nusa Tenggara Province, the expansion of beach resort tourism has led to the extraction of massive amounts of water from aquifers that are under stress, thus compromising local residents' access to freshwater; water shortages place a disproportional impact on women.
Costa Rica (Central America) Chapter 4	Water demand of the rapidly growing resort and residential tourism sector in Guanacaste Province is inadequately recorded, but water extraction rates are increasing and have led to claims of water scarcity by residents in some places. In the Sardinal district, salinisation of aquifers has been reported, due to overexploitation by hotels.
India (South Asia) Chapter 9	Greenfield airport expansion in the southwestern Indian state Goa has led to the extraction of thousands of trees and the diversion of agricultural irrigation water to feed the airport's water demand. The construction work has removed reserved forests, infringed on wetlands and water bodies and destroyed archaeologically and religiously important sites.

Air, water and plastic pollution are also major features of tourism's extractive practices, with detrimental impacts on affected people's health and wellbeing. In many coastal tourist destinations in the Global South, hotels dump their wastewater directly into the ocean, raising concerns about human health and the integrity of coastal ecosystems (e.g. Gössling, 2003; Wilkinson, 2003; Ong, Storey and Minnery, 2011). In the Maldives, an artificial island – Thilafushi – has been created to contain all the waste from the country's holiday islands.

Erasure

The final practice of dispossession discussed here and illustrated by a set of examples from the case studies is erasure. This practice is often at play when a dominant ethnic or political group in society or an occupying force instrumentalises the tourism sector for purposes of cultural domination, annihilation of specific cultural and religious practices or appropriation of cultural artefacts and archaeologically important places. Such processes of erasure through tourism are exemplified by several examples in Table 10.4.

Erasure can be a sudden or forceful event, e.g. when a tourist resort is built on the burial grounds of an indigenous group, but it can also be a slow process of cultural erosion. The continuous foreignisation of space that often occurs in the wake of residential tourism development is one example for a gradual erasure of local culture. Some tourism stakeholders may appropriate minority and Indigenous cultures through some form of 'Disneyfication' or 'museumification', where certain cultural elements (e.g. exotic dances or traditional clothes) are displayed for tourism purposes, while other elements (e.g. certain spiritual practices) are denied or erased.

Table 10.4 Practices of dispossession in tourism – Erasure

Case study	Characteristics
Thailand (Southeast Asia) Chapter 5	Post-tsunami tourism development in coastal areas and consolidation of settlements of indigenous, formerly sea-faring people in marine protected areas in the Andaman region of southern Thailand have erased Indigenous practices, such as traditional boat construction, commodified culturally important artefacts, and prevented local communities from accessing their ceremonial places and spirit shrines.
Honduras (Central America) Chapter 5	Post-disaster and post-conflict tourism development and protected area expansion along the North Coast threatens to erase communal land rights and socio-cultural practices of Garifuna communities whose distinct cultural traditions, including language, dance and music, were declared a Masterpiece of the Oral and Intangible Heritage of Humanity by UNESCO.
Myanmar (Southeast Asia) Chapter 6	In Bagan, a world-renowned cultural landscape in Myanmar with hundreds of ancient temples and pagodas, the country's military regime removed traditional caretakers of cultural heritage sites in the early 1990s, thereby erasing significant repositories of living historic and cultural knowledge.
Sri Lanka (South Asia) Chapter 6	Sri Lanka's military has confiscated large tracts of land from the Tamil minority population during and after the long civil war. It has also transformed sites of victory over the Liberation Tigers of Tamil Eelam into monument sites and places of triumphalism that tell a one-sided story of the conflict and risk to erase Tamil cultural and collective memory in a process of 'Sinhalisation'.
Palestine (Middle East) Chapter 6	In occupied East Jerusalem, the Israeli government has not only appropriated important historical sites from the Palestinians but also attempted over the past two decades to expand archaeological-touristic parks that infringe on Palestinian residential areas. Tourism authorities tried to remove Muslim places of interest from tourist maps.
Mozambique (Southern Africa) Chapter 7	The establishment of the Limpopo National Park called for the 'voluntary' resettlement of thousands of people from areas with the highest tourism potential in the park's core zone. Authorities failed to consider less tangible losses, such as leaving cemeteries behind or losing access to traditional plants and animals, thus erasing cultural memory and Indigenous knowledge.

A final type of erasure that is worth mentioning in this context is the deagrarianisation and depeasantisation of communities through tourism-related land and resource grabs. Agroecological practices and knowledges in Indigenous and non-indigenous communities in the Global South are often deeply embedded in local culture and religious beliefs. Tourism-induced dispossession and the physical or economic displacement of farmers, fisherfolk, huntergatherer communities and pastoralists risk erasing invaluable Indigenous and traditional knowledge in the domains of agriculture, fisheries and forestry and ultimately disconnect rural and coastal communities from land and natural resources. In the long-term, this can have major implications on food security and social stability not only at the local but also at the national level.

Resistance

Resistance to tourism-related land grabs and displacement has rarely been systematically analysed and conceptualised, despite a relatively large body of literature on resistance to land grabbing for other purposes, such as plantation agriculture and mining (for an overview, see Hall et al., 2015). An exception is the conceptualisation and empirical examination of resistance movements against mass tourism and foreignisation of space in Goa undertaken by Routledge (2001) and Saldanha (2002), although their studies did not include any micro-level analysis of land grabs and displacement at the community level.

As most case studies discussed in Chapters 3–9 have shown, tourism-related land grabs and displacement have been met with a range of resistance strategies deployed by affected local communities and their allies. Most common are *spontaneous, defensive, locally based and open forms of resistance*, such as tearing down fences of enclosed tourism facilities (cf. the case of Rote Ndao in Eastern Indonesia in Chapter 4), interruption of road traffic (cf. the case of Haiti's Île à Vâche in Chapter 5) and violent clashes with construction company workers, hotel guards and security personnel (cf. the case of the Urak Lawoi Indigenous community in Thailand in Chapter 5). Setting up protest camps and occupying tourism zones that have infringed on customary property are other strategies that have been devised by defiant local communities (cf. the case of Koh Kong Province in Cambodia in Chapter 3 and the case of Boracay and Sicogon Islands in the Philippines in Chapters 3 and 5 respectively). Yet, while collective and overt defiance by community groups has drawn a certain degree of attention from the local and international media, such resistance movements are often short-lived and easily squashed by brute government force or violent actions by security personnel of tourism operators. Similarly, *everyday forms of local resistance* (cf. Scott, 1986; Kerkvliet, 2009), for instance in the form of non-compliance, low-key obstructions and covert counteractions, have been largely ineffective or – at best – prolonged the processes of dispossession and displacement (e.g. Neef and Touch, 2015). Hence, place-based and reactive responses by small groups of protesters seem to conform with Harvey's (1996) notion of 'parochialist politics', referring to locally embedded and issue-specific movements that remain "relatively powerless when confronted by capital's global reach" (Routledge, 2001, p. 235).

The success of resistance against tourism-induced land grabs and displacements seems to depend less on the actual practices of resistance but rather on (1) the assemblage of resistance, i.e. the particular coalitions that protesters are able to build at various levels, and (2) resistance identities that can lead to the formation of collective resistance movements (cf. Routledge, 2001; Castells, 2009). The fight of the formerly seafaring Urak Lawoi against a Bangkok-based tourism investor in southern Phuket, Thailand, was partially successful because local community was able to forge a broad coalition with a national non-governmental organisation (NGO), domestic and foreign academics, and the Thai Department of Special Investigation (Neef et al., 2018; Chapter 5). It was also helped by developing a resistance identity as Indigenous settlers defending the last communal beach on the island and by making reference to a visit of the late Thai king to the area. Similarly, the ForBALI association that was successful in preventing the Benoa Bay Reclamation megaproject on the island of Bali, Indonesia, brought together environmental activists, village leaders, politicians, journalists, academics, students, and artists, thereby creating a mass movement that evoked a strong sense of Balinese identity (Benge and Neef, 2018; Chapter 4). The association organised a series of protest marches, public fora and popular concerts to express a strong civil society stance against external (Javanese) investors.

By working alongside domestic or international NGOs in their struggle against predatory tourism investors, local communities can build capacity, raise their own confidence level and strengthen their legal literacy. Such forms of *advocacy resistance* have been successful to a certain degree in Goa, India, where the local NGO scene is particularly diverse and vivid with a long-standing history that dates back to the mid-1970s (Saldanha, 2002). Community-based tourism initiatives and forest concessions in Guatemala's Maya Biosphere Reserve have benefited from their alliance with various international environmental and human rights NGOs in their fight against foreign and domestic promoters of large-scale archaeotourism (Chapter 8).

Gendered and generational dynamics of responses to land grabbing and displacement have received relatively modest scholarly attention in the wider literature on the global rush for land (Hall et al., 2015). Studies have remained even more scant in the field of tourism-related land grabbing. Reports on the gendered dimension of resistance remain sketchy and anecdotal. The reportedly successful lobbying of the Tanzanian government by hundreds of Indigenous Maasai pastoral women against large-scale evictions from wildlife parks is one such case (IWGIA, 2019; Chapter 7). Women have also played a leading role in protests against a Chinese state-owned corporation in Cambodia's Koh Kong Province (Chapter 3). Resistance movements in other case studies have been more masculinist, e.g. in the case of the Indigenous struggles against the 'Maya Train' megaproject in Mexico (Chapter 9).

It has been alleged that younger generations – who would be enticed by presumed job opportunities in the tourism sector – have more positive attitudes towards local tourism development than older people who are deemed to be more inclined to maintaining the status quo and defending customary ownership of land due to their stronger place attachment. Yet this may well be a sweeping generalisation, as the case of the Mexican megaresort of Cancún shows, where more than a hundred local children promoted an injunction against the destruction of a mangrove wetland by tourism investors by invoking their right to a healthy environment (Chapter 3). Protests against tourism-related land grabs in post-earthquake Haiti (Chapter 5) and on the island of Boracay in the Philippines (Chapter 3) have been led by young, well-educated leaders. By contrast, local resistance against tourism-related relocation and enclosure in the Indigenous Moken community of Baan Tungwa on Thailand's Andaman Coast has been organised by an elderly community leader (Chapter 5).

A final point that is important to make is that in many countries in the Global South, defenders of land rights against corporate and government tourism investors are putting their own freedom and even their life at risk. Forty per cent of the more than 300 human rights defenders who were killed globally in 2019 – as reported to the International Human Rights Defenders Memorial – worked on land rights, Indigenous people's rights and environmental rights, with land rights defenders in Colombia, the Philippines, Honduras, Brazil, Guatemala and India being most at risk of being killed (Frontline Defenders, 2020). In Myanmar, one third of political prisoners in 2015 were land rights activists. Hundreds of Indigenous human rights defenders and Indigenous villagers – many of whom had resisted military-backed tourism projects – have been arrested or detained by government forces in Bangladesh (Kapaeeng Foundation, 2018). In Cambodia, authoritarian populism and crackdowns on media and NGOs have considerably constrained spaces for social movements and human rights advocacy (e.g. Beban, Schoenberger and Lamb, 2020).

Concluding remarks

Drawing on Harvey's (2006) mechanisms of 'accumulation by dispossession' and adapting Devine and Ojeda's (2017) and Neef et al.'s (2018) practices of dispossession, this chapter has synthesised the findings from the 31 case studies presented in Chapters 3–9. This synthesis has unpacked the patterns in which communities in the Global South have been impacted by tourism-induced dispossession and displacement. While eviction, enclosure, extraction and erasure come in various shapes and forms, they all have detrimental impacts on the rights of communities and individuals in tourism destinations. The chapter has also discussed various forms of resistance that communities and civil society groups have employed in the struggle against tourism-related land grabs and displacement. It has highlighted the fact that land rights defenders operate under high risks to their freedom and life, even in the more democratic societies of the Global South.

The next chapter will introduce a range of instruments and guidelines for land governance that can be deployed to protect communities from tourism-related dispossession and displacement. Both their potential and shortcomings will be highlighted.

References

Beban, A., Schoenberger, L. and Lamb, V. (2020) 'Pockets of liberal media in authoritarian regimes: what the crackdown on emancipatory spaces means for rural social movements in Cambodia', *The Journal of Peasant Studies*, 47 (1): 95–115.

Becken, S. (2014) 'Water equity – Contrasting tourism water use with that of the local community', *Water Resources and Industry*, 7–8: 9–22.

Benge, L. and Neef, A. (2018) 'Tourism in Bali at the interface of resource conflicts, water crisis, and security threats' in Neef, A. and Grayman, J.H. (eds) *The Tourism-Disaster-Conflict Nexus*. Emerald Publishing: Bingley, pp. 33–52.

Castells, M. (2009) *The Power of Identity* (2nd edition). Wiley-Blackwell: Chichester.

Cole, S.K.G., Cañada Mullor, E., Ma, Y. and Sandang, Y (2020) '"Tourism, water, and gender"—An international review of an unexplored nexus', *WIREs Water*, 7: e1442. https://doi.org/10.1002/wat2.1442.

De Soto, H. (2000) *The Mystery of Capital: Why Capitalism Triumphs in the West and Fails Everywhere Else*. Basic Books: New York.

Devine, J. and Ojeda, D. (2017) 'Violence and dispossession in tourism development: A critical geographical approach', *Journal of Sustainable Tourism*, 25 (5): 605–617.

Frontline Defenders (2020) *Frontline Defenders Global Analysis 2019*. Frontline Defenders: Dublin, Brussels. Available at: www.frontlinedefenders.org/sites/default/files/global_analysis_2019_web.pdf (accessed 12 August 2020).

Gibson, C. (2019) 'Critical tourism studies: New directions for volatile times', *Tourism Geographies*. doi:10.1080/14616688.2019.1647453.

Gössling, S. (2003) 'Tourism and development in tropical islands: Political ecology perspectives' in Gössling, S. (ed.) *Tourism and Development in Tropical Islands. Political Ecology Perspectives*. Edward Elgar: Cheltenham, Northampton, MA, pp. 1–37.

Gössling, S., Peeters, P., Hall, M.C., Ceron, J., Dubois, G. and Lehmann, L. (2012) 'Tourism and water use: Supply, demand, and security. An international review', *Tourism Management*, 33: 1–15.

Hall, R., Edelman, M., Borras, S.M. Jr., Scoones, I., White, B. and Wolford, W. (2015) 'Resistance, acquiescence or incorporation? An introduction to land grabbing and political reactions "from below"', *Journal of Peasant Studies*, 42: 3–4.

Harvey, D. (1996) *Justice, Nature and the Geography of Difference*. Blackwell: Oxford.

Harvey, D. (2006) 'Neo-liberalism as creative destruction', *Geografiska Annaler*, 88B (2): 145–158.

IWGIA (2019) *Annual Report 2019*. International Work Group for Indigenous Affairs (IWGIA): Copenhagen.

Kapaeeng Foundation (2018) 'Government plans to acquire 700 acres of land for establishment of Special Tourism Zone at Alutila hills in Khagrachari which leads to eviction of hundreds indigenous families'. Available at: www.kapaeeng.org/governm ent-plans-to-acquire-700-acres-of-land-for-establishment-of-special-tourism-zone-at

-alutila-hills-in-khagrachari-which-leads-to-eviction-of-hundreds-indigenous-families/ (accessed 12 January 2019).

Kerkvliet, B.J.T. (2009) 'Everyday politics in peasant societies (and ours)', *The Journal of Peasant Studies*, 36 (1): 227–243.

Neef, A. and Touch, S. (2015) 'Local responses to land grabbing and displacement in rural Cambodia' in Price, S. and Singer, J. (eds) *Global Implications of Development, Climate Change and Disasters: Responses to Displacement from Asia–Pacific*. Routledge: London, New York, pp. 124–141.

Neef, A., Attavanich, M., Kongpan, P. and Jongkraichak, M. (2018) 'Tsunami, tourism and threats to local livelihoods: The case of indigenous sea nomads in southern Thailand' in Neef, A. and Grayman, J.H. (eds) *The Tourism-Disaster-Conflict Nexus*. Emerald Publishing: Bingley, pp. 141–164.

OHCHR (1997) *General Comment No. 7: The Right to Adequate Housing (Art. 11 (1) of the Covenant): Forced Evictions*. United Nations Human Rights Office of the High Commissioner (OHCHR). Available at: https://tbinternet.ohchr.org/_layouts/treatybody external/Download.aspx?symbolno=INT/CESCR/GEC/6430&Lang=en (accessed 24 January 2019).

OHCHR (2018) *Basic Principles and Guidelines on Development-Based Evictions and Displacement*. Available at: www.ohchr.org/Documents/Issues/Housing/Guidelines_en.pdf (accessed 26 October 2020).

Ong, L.T.J., Storey, D. and Minnery, J. (2011) 'Beyond the beach: balancing environmental and socio-cultural sustainability in Boracay, the Philippines', *Tourism Geographies*, 13 (4): 549–569.

Routledge, P. (2001) '"Selling the rain", resisting the sale: Resistant identities and the conflict over tourism in Goa', *Social & Cultural Geography*, 2 (2): 221–240.

Saarinen, J. and Wall-Reinius, S. (2019) 'Enclaves in tourism: Producing and governing exclusive spaces for tourism', *Tourism Geographies*, 21 (5): 739–748.

Saldanha, A. (2002) 'Identity, spatiality and post-colonial resistance: Geographies of the tourism critique in Goa', *Current Issues in Tourism*, 5 (2): 94–111.

Scheyvens, R. (2002) *Tourism for Development: Empowering Communities*. Pearson Education Limited: London.

Scott, J. (1986) 'Everyday forms of peasant resistance', *Journal of Peasant Studies*, 13 (2): 5–35.

Tourism Concern (2012) *Water Equity in Tourism – A Human Right, A Global Responsibility*. Available at: www.tourismconcern.org.uk/wp-content/uploads/2014/10/Water-Equity-Tourism-Report-TC.pdf (accessed 22 January 2019).

Wilkinson, P.F. (2003) 'Tourism policy and planning in St. Lucia' in Gössling, S. (ed.) *Tourism and Development in Tropical Islands*. Political Ecology Perspectives. Edward Elgar: Cheltenham, Northampton, MA, pp. 88–120.

11 Instruments and guidelines for land governance and protection from dispossession and displacement

Potential applications in the field of tourism

There are a range of laws and regulations pertaining to the operational aspects of the global tourism industry albeit with different levels of local, national and international enforcement (Swarbrooke, 1999; Mowforth and Munt, 2016). There are much fewer regulations that are related to the responsibilities of tourists (Mason, 2016). Cosmopolitan tourists enjoy a range of rights, including the right to free and unrestricted movement, the right not to be exploited by local businesses and individuals, the right to be secure while travelling, and even the right to a safe and clean environment (Swarbrooke, 1999). These rights are to be protected by host communities, government agencies and the tourism industry. Yet the high level of protection and enjoyment of these rights often stands in stark contrast to the more precarious rights of local communities and individuals that may be compromised by compulsory land acquisition for tourism infrastructure, corporate land grabbing, enclosure of essential natural resources, and physical and economic displacement.

This chapter looks at existing frameworks and guiding principles that have the potential to strengthen land and resource rights and protect both Indigenous and non-indigenous communities from tourism-induced land grabs and displacement. While the first three sections examine international human rights law and the particular rights of Indigenous and tribal peoples, the final four sections explore the potential of voluntary guidelines, environmental and social safeguards of international financial institutions and corporate codes of conduct for controlling tourism-induced land grabbing, involuntary resettlement and planned relocation.

Tourism-related land grabs and international human rights law

Tourism is linked to a wide range of human rights. Several human rights principles as enshrined in international human rights frameworks pertain to land and other natural resources. Tourism-related land grabs directly affect these rights as presented in Box 11.1. There are also various human rights that may be indirectly impacted by tourism-related land grabbing. Forced eviction and involuntary resettlement, for instance, often have adverse effects on health and self-determination, affect the right to non-interference with privacy, family and home, and may expose evicted or resettled people to economic and sexual exploitation after the relocation.

Box 11.1 Key human rights principles and issues in tourism

Directly related to land and other natural resources:

- The right to own property, including land
- The right to adequate housing
- The right to protection from forced displacement
- The right to food
- The right to water and sanitation
- The rights of indigenous peoples

Indirectly related to land and other natural resources:

- The right to life and health
- The right to dignity and privacy
- The right to protection from economic and cultural exploitation
- The right to participation and self-determination
- The right to be protected from child labour and sexual exploitation

Sources: OHCHR, 1997; Tourism Concern, 2014;
Roundtable Human Rights in Tourism, 2016

A major gap in international human rights law remain the rights of future generations. These are impacted when land is permanently removed from customary, legitimate right-holders and even in land lease arrangements that may extend to 99 years in some locations, thereby compromising the rights of three to four generations.

The right to property, adequate housing and protection from displacement

The *right to property* has remained somewhat controversial among international human rights lawyers. Article 17 of the Universal Declaration of Human Rights (UDHR) holds that 'no one shall be arbitrarily deprived of his property' and Article 25(1) of the UDHR guarantees the *right to housing* and *prohibits forced evictions*. Both articles can be interpreted as an international and legally binding recognition that land grabbing, including for tourism purposes, is a violation of these basic human rights. The UN Guiding Principles on Internal Displacement (UNGPID) build on these human rights principles and maintain that the prohibition of arbitrary displacement includes "large-scale development projects, which are not justified by compelling and overriding public interests" (Principle 6.2.c). They further hold that "prior to any decision requiring the displacement of persons, the authorities concerned shall ensure that all feasible alternatives are explored in order to avoid displacement altogether" (Principle 7.1). The UNGPID also make provisions for "full information on the reasons and procedures for [people's] displacement" and call for "the free and informed consent of those to be displaced" (Principle 7.3.b/c). They also urge authorities to "involve those affected, particularly women, in the planning and management of their relocation" (Principle 7.3.d). Principle 9 is particularly pertinent for tourism- and conservation-related displacement as it recognises the particular obligation of states "to protect against the displacement of indigenous peoples, minorities, peasants, pastoralists and other groups with a special dependency on and attachment to their lands" (Principle 9). Remarkably, the UNGPID also provide for internally displaced persons' "right of access to the grave sites of their deceased relatives" (Principle 16.4), which is a right that has often been ignored in tourism- and conservation-related evictions and resettlements. Two problems arise in the application of UNGPID to the tourism sector: first, the principles – while directly drawing on international human rights law – are not legally binding, and, second, tourism tends to feature less prominently in the international debate on the rights of 'internally displaced people' than traditional extractive industries or war and conflict.

Right to food, water and sanitation

The *right to food* is enshrined in the 1948 Universal Declaration of Human Rights (UDHR) and the International Covenant on Economic, Social and Cultural Rights (ICESCR) which explicitly recognises 'the fundamental right of everyone to be free from hunger' (Art. 11(2)). The ICESCR obligates states to take appropriate measures, either in their individual capacity or through international partnerships, to address all forms of food insecurity (Dhanarajan, 2015). In many of the case studies presented in Chapters 3–9 it has become evident that most tourism-related land grabs had immediate and adverse impacts on the food security (and the food sovereignty) of evicted and resettled people. Yet, the advocates of tourism development often argue that food insecurity will only be transitional and that tourism projects, conservation zones or major infrastructure projects (like airports or railways) provide future income opportunities for displaced communities. Unfortunately, several of the case studies presented in this report and many other studies have shown that such hopes and promises have rarely materialised for the direct victims of development-induced displacement (cf. Price, 2015).

The human *right to water and sanitation* as a prerequisite for leading a life in human dignity and for the realisation of various other human rights has been explicitly recognised through the legally binding UN Resolution 64/292 (28 July 2010). In the context of tourism-related land grabbing, violations of this fundamental human right may stem from: (1) diversion of freshwater supplies to tourism businesses at the expense of providing drinking and irrigation water to communities and other stakeholders, such as small-scale farmers, fisherfolks or pastoralists; (2) contamination of freshwater sources through pesticides, e.g. from golf courses; and (3) evictions of communities from areas with adequate water supply. Several examples of how the human right to water has been disrespected by various stakeholders in the tourism sector's supply chain have been discussed in the previous chapters. In the future, it will be an important task of human rights advocates and international human rights lawyers to acknowledge and raise awareness that the tourism sector is not only an extractive industry with an enormous water footprint, but that many tourism stakeholders are also actively engaged in grabbing water resources at the expense of local communities' enjoyment of this fundamental human right.

If governments take their human rights obligations under international legal frameworks seriously, they need to implement their fundamental duty to protect the rights of their citizens to land, property, housing and access to water and sanitation for essential livelihood needs. This includes protection against infringements by tourism businesses, tourism infrastructure and tourism zoning.

The ILO's Indigenous and Tribal Peoples Convention

The 1989 Convention (No. 169) *Concerning Indigenous and Tribal Peoples in Independent Countries of the International Labor Organization* (commonly referred to as ILO 169) was written to pressure governments to enact special legislation for the rights of 'Indigenous and tribal peoples' to land, bilingual education, political and economic autonomy, and fair labour practices. ILO 169's definition of the groups protected by the convention is relatively loose, applying to tribal peoples, Indigenous peoples, and peoples present prior to colonisation who have continued to retain 'traditional' cultural institutions (ILO, n.d.).

Land rights are central to the ILO 169, since without control over territory, Indigenous and tribal peoples have no control over their own development. ILO 169 also recognises that land is tied to maintaining cultural identity. The second part of the convention (Articles 13–19) deals explicitly with Indigenous land rights. Article 13 declares that governments need to respect the cultural and spiritual value that Indigenous peoples attach to their lands, territories, or both and, in particular, the collective nature of that relationship. The remaining articles make it unequivocally clear that Indigenous peoples are to be afforded their rights not just to land occupied by them, but also to areas that they had traditionally accessed for subsistence and other activities. Governments are required to safeguard and guarantee the protection of Indigenous rights to ownership and are called upon to adopt "adequate procedures within the national legal system to resolve land claims by the peoples concerned" (ILO, n.d.).

ILO 169 is potentially a powerful legal framework to protect Indigenous and tribal peoples from tourism-induced land grabs. Yet it suffers from very low uptake by the international community, with only 23 nation states having ratified the convention by October 2020. Most of the signatories to ILO 169 hail from Latin America, a region with the highest potential for applying the convention. When applied to the tourism industry, it requires any tourism stakeholder that wants to acquire land – including government bodies – to pre-obtain the free, prior and informed consent (FPIC) of local communities. However, some tourism watchdogs have argued that a major weakness of ILO 169 is that injured parties, including Indigenous peoples, have no right to veto land acquisitions (IHRB and Tourism Concern, 2012).

The UN Declaration on the Rights of Indigenous Peoples (UNDRIP)

The UN Declaration on the Rights of Indigenous Peoples (UNDRIP) was negotiated between nation states and Indigenous peoples over a period of 20 years before being adopted by the UN General Assembly in 2007. The Declaration features 46 articles covering a wide range of human rights issues (United Nations, 2008). By 2016, nearly all UN member states supported the Declaration, following initial opposition by several settler colonial states, including the USA, Canada, New Zealand and Australia. The UNDRIP has been hailed as a landmark achievement by Indigenous peoples in obtaining international recognition of their rights to self-determination and right to land and other natural resource rights. While building on international human rights law, as a General Assembly Declaration the UNDRIP is not a legally binding instrument nor does it create a new set of rights for Indigenous peoples. Rather, the declaration builds on existing human rights standards and applies them to the particular situation of Indigenous peoples, with emphasis on the recognition of collective rights to Indigenous territories (see selected articles in Box 11.2).

Box 11.2 UNDRIP articles with particular relevance to the protection of Indigenous peoples' collective rights to land and other natural resources

Article 10

Indigenous peoples shall not be forcibly removed from their lands or territories. No relocation shall take place without the free, prior and informed consent of the indigenous peoples concerned and after agreement on just and fair compensation and, where possible, with the option of return.

Article 25

Indigenous peoples have the right to maintain and strengthen their distinctive spiritual relationship with their traditionally owned or otherwise occupied and used lands, territories, waters and coastal seas and other resources and to uphold their responsibilities to future generations in this regard.

Article 26

1   Indigenous peoples have the right to the lands, territories and resources which they have traditionally owned, occupied or otherwise used or acquired.
2   [...]
3   States shall give legal recognition and protection to these lands, territories and resources.

Source: United Nations, 2008

The UNDRIP provides nation states with a comprehensive framework to reduce inequality and provide remediation when Indigenous peoples' rights have been violated. A major strength of the Declaration as a tool of protection from tourism-related land grabs is its denouncement of all forms of forced relocation of Indigenous peoples, emphasising the explicit need for FPIC and just and fair compensation. As most governments have ratified the UNDRIP, they are legally bound to acknowledge, respect and formalise Indigenous land rights in ways that are inclusive of local conceptions of resource ownership, including such notions as the inseparability of people and land among many Indigenous groups. Yet, the document does not define the term 'Indigenous peoples', which limits its application in contexts, where governments do not acknowledge the concept of indigeneity for their own Indigenous groups and have developed alternative classifications, such as ethnic minorities, hilltribes, ethnic nationalities or scheduled tribes.

Building on the UNDRIP, the Pacific Asia Travel Alliance (PATA) in association with the World Indigenous Tourism Association (WINTA) have developed their own set of principles (the Larrakia Declaration) for how Indigenous peoples want to engage with the tourism sector. The first two principles make explicit reference to Indigenous rights to land and natural resources (Box 11.3).

Box 11.3 The six principles of the 2012 Larrakia Declaration

1   Respect for customary law and lore, land and water, traditional knowledge, traditional cultural expressions, cultural heritage that will underpin all tourism decisions.
2   Indigenous culture and the land and waters on which it is based, will be protected and promoted through well-managed tourism practices and appropriate interpretation.
3   Indigenous peoples will determine the extent and nature and organizational arrangements for their participation in tourism and that governments and multilateral agencies will support the empowerment of Indigenous people.

4   Governments have a duty to consult and accommodate Indigenous peoples before undertaking decisions on public policy and programs designed to foster the development of Indigenous tourism.

5   The tourism industry will respect Indigenous intellectual property rights, cultures and traditional practices, the need for sustainable and equitable business partnerships and the proper care of the environment and communities that support them.

6   Equitable partnerships between the tourism industry and Indigenous people will include the sharing of cultural awareness and skills development which support the wellbeing of communities and enable enhancement of individual livelihoods.

Source: PATA and WINTA, 2015, p. 13

While it is important to acknowledge the particular need for protection of Indigenous peoples' land in the context of tourism development, it needs to be recognised that many non-indigenous communities also have a deep connection with and dependency on their land and other communally shared resources and may be similarly at risk of displacement from tourism-related land grabs, as has been shown in several examples in this study. The following two subsections look at two sets of voluntary guidelines that are broader in scope and address the vulnerabilities of both Indigenous and non-indigenous populations.

The FAO's Voluntary Guidelines and their implications for tourism-related land grabs

The Voluntary Guidelines on the Responsible Governance of Tenure of Land, Fisheries and Forests in the Context of National Food Security (VGGT) were drawn up by the Food and Agricultural Organization of the United Nations (FAO) and adopted by the Committee on World Food Security (CFS) in 2012. Their aim is to achieve 'food security for all' and to promote 'responsible governance of tenure of land, fisheries and forests, with respect to all forms of tenure: public, private, communal, indigenous, customary, and informal' (FAO, 2012). These aims are grounded in fundamental human rights frameworks.

What makes the VGGT distinct from previously established international land governance frameworks is that they call for the recognition and protection of 'legitimate' tenure rights, which include all forms of customary, informal and subsidiary rights, even if they are not (yet) acknowledged and protected by statutory law (FAO, 2015; Hall and Scoones, 2016). This recognition is also extended to publicly owned lands, which is particularly significant for tourism zone development and large-scale tourism infrastructure projects (such as airport constructions and expansions, discussed in Chapter 9), where public land may overlap with customary rights of long-established communities that tend to be delegitimised by government actors and project developers (see Chapter 3). Adhering to the VGGT would mean that hotels, tour operators and other tourism businesses need to respect the legitimate rights to land and other natural resources of all stakeholder groups (e.g. Indigenous peoples, local communities, customary landowners), even when these rights are not codified by national legislation.

The VGGT also make explicit reference to pastures, fishing grounds and forests that are used and managed communally rather than by individual owners or custodians. Such 'commons' tend to be targeted by many private tourism investors, as communal rights are often not legally recognised by national governments or given much less priority in legislative frameworks than private and state property. Communally owned and managed pastures and forests are also at high risk of expropriation when states aim to expand their protected areas for both wildlife conservation and eco-tourism, as exemplified in Chapter 7. Governments that take the VGGT as a guiding framework should refrain from prioritising tourism development over local livelihoods derived from agriculture, fisheries and forestry.

Finally, the VGGT call for redistributive reforms to provide wider and fairer access to resources for men and women equally. This principle is of relevance for regions and countries with a particularly skewed distribution of land and other natural resources, often as a consequence of their colonial history, such as in Asia (e.g. Philippines, Indonesia and India), several Latin American countries (e.g. Honduras, Peru and Brazil) and a number of African nations (e.g. Tanzania, Mozambique and South Africa). In many of these countries, unfettered tourism development and expansion of protected areas have exacerbated historically determined distributional inequity, while communities and civil society groups have advocated for land reform, redistribution and restitution for decades.

A major shortcoming of the VGGT is the absence of provisions for water rights which has been criticised by development practitioners, academics and civil society groups alike (e.g. Brüntrup et al., 2014; Paoloni and Onorati, 2014). This is particularly problematic for the resort tourism sector, whose water demand is often several times higher than other sectors (cf. Chapter 4), possibly with the exception of farming. Another limitation is that the principle of Free, Prior and Informed Consent (FPIC) – which is explicitly addressed in relation to Indigenous peoples in the VGGT with reference to the UNDRIP – is not extended to non-indigenous social groups (Paoloni and Onorati, 2014). This poses challenges for human rights groups that advocate for FPIC among non-indigenous communities (for example, in the context of urban and peri-urban land grabs, e.g. for sports mega-events and urban tourist attractions, as outlined in Chapter 8), but also for 'Indigenous peoples' that are not officially recognised as such by their governments, for instance in Tanzania, Laos, Indonesia, Thailand and India.

Another shortcoming of the VGGT is that they do not categorically oppose controversial land deals and leave room for diverging interpretations of what constitutes an illegitimate or unethical land acquisition or lease (Neef, 2016).

Consequently, governments, investors and bilateral or international agencies can make reference to certain elements in the text to justify their focus on national economic growth and to emphasise the public benefits of land deals (Neef, 2019). This is particularly common in national and international tourism discourses, where the tourism sector continues to be presented as an inherently benign, non-extractive and even peace-making industry (see Chapter 1).

A promising development is that the United Kingdom, the USA, Germany, France, the African Union Land Policy Initiative and the Food and Agricultural Organization of the United Nations (FAO) have jointly developed a land investment due diligence framework based on the VGGT and other international standards, to guide private sector investments under their New Alliance for Food Security and Nutrition. Yet, while the VGGT are gradually being adopted by an increasing number of stakeholders at government, donor and corporate levels, they remain a soft law instrument that can support a range of human rights advocacy measures and raise further awareness among various stakeholders, but do not enable the victims of tourism-related dispossession and displacement to take legal action against governments and corporations.

The UN Guiding Principles on Business and Human Rights and their relevance to land acquisitions by the tourism industry

The UN Guiding Principles on Business and Human Rights (UNGP) were adopted by the UN Human Rights Council in 2011 and are arguably the most authoritative and internationally recognised framework for business and human rights, since they are backed by UN member states and are the outcome of extensive consultations with many stakeholders over a period of six years. The UNGP have also been referred to as the UN's 'Protect, Respect, Remedy Framework'. These three pillars cover the state's duty to *protect* against human rights abuses, the corporate responsibility to *respect* the human rights of all peoples, and the contractual parties' obligation to ensure access to effective *remedy* when protection fails (cf. Box 11.4).

While the UNGP do not create obligations under international law, they are increasingly becoming a concrete reference framework on the obligations of states and the responsibilities of businesses to respect human rights in relation to tourism. The application of the UNGP is particularly pertinent when businesses operate in countries where adherence to international human rights norms and standards in the tourism sector are weak due to lack of government will, capacity and/or resources, or because of on-going or recent violent conflict, such as in Sri Lanka, Palestine, Bangladesh and Honduras.

Box 11.4 The three pillars of the Guiding Principles on Business and Human Rights

1   The state's duty to *protect* against human rights abuses: States must guarantee protection against human rights violations committed by

third parties, such as businesses, within their territory. This calls for appropriate measures to prevent, investigate, prosecute and compensate for human rights violations.

2 The corporate responsibility to *respect* human rights: Businesses should respect human rights and avoid negative impacts that are caused directly or through their business relations. In order to assume responsibility, businesses should possess corresponding principles and procedures and act with due diligence.

3 The contractual parties' obligation to ensure access to effective *remedy* when protection fails: States must take adequate measures to provide access to an effective remedy and appropriate compensation for the affected parties. In addition to judicial mechanisms, states must also provide non-judicial grievance mechanisms. Moreover, businesses should also provide effective grievance mechanisms at an operative level, or participate in such mechanisms.

Source: OHCHR, 2011

Nevertheless, the implementation of the UNGP by nation states has proven difficult. The EU Commission has asked its member state governments to prepare action plans for the implementation of UNGP, and 15 of its 27 members have produced this as of October 2020. Internationally, only nine other countries (UK, Norway, Columbia, Switzerland, USA, Chile, Kenya, Thailand and Japan) have developed their own national action plan, while a range of other countries, including in Africa, Asia and Latin America, have committed to draft an action plan or have taken first steps in the development of such a plan. The UK's and France's national action plans both make specific reference to large-scale land acquisitions, but the tourism sector is not mentioned in either of them. Hence, while the UNGP are gradually being incorporated into government policies and hard law, it is obvious that tourism to date has escaped the same levels of human rights scrutiny as other sectors (IHRB and Tourism Concern, 2012). It will be important to put tourism on the agenda of the UNGP implementation process and emphasise the responsibility of tourism businesses to respect land rights of Indigenous and non-indigenous peoples as essential human rights.

A major weakness of the UNGP is that they do not develop duties for individual states to monitor and regulate the human rights impacts of their home business enterprises beyond national borders (i.e. extraterritorial obligations – ETOs). Hence, foreign tourism companies can easily a adopt less stringent social and environmental standards in host countries, without being subjected to any punitive measures from their home governments. This is unsatisfactory as ETOs are internationally recognised in other areas, such as sex tourism (Van Huijstee, Ricco and Ceresna-Chaturvedi, 2012). There are, however, a number of encouraging developments, such as a study commissioned by the European Parliament's Sub-Committee on Human Rights which focuses on access to legal remedies for victims of corporate human rights abuses in third countries. For tourism businesses, this would mean that they must support appropriate remedial action for all social groups that have been adversely affected by their activities, e.g. through land restitution measures and adequate compensation payments as required by national or international legal frameworks and voluntary guidelines.

There are various other codes of conduct that are closely related to UNDG. More than 5,000 global corporations have joined the UN Global Compact, which calls on businesses to "support and respect the protection of internationally proclaimed human rights" (Principle 1), "make sure that they are not complicit in human rights abuses" (Principle 2) and "support a precautionary approach to environmental challenges" (Principle 7). Several multinational hotel chains, such as InterContinental Hotels, Hyatt and Shangri-La, have also started to explicitly recognise their human rights obligations and have implemented measures such as human rights policies, reporting on human rights issues, providing ethics training for their staff and signing up to the UN Global Compact (Bauer, 2015). Human rights policies and self-regulation measures developed and implemented by tourism operators and encouraged by global lobby organisations, such as the World Travel and Tourism Council (WTTC) and the World Travel Organization (UNWTO), focus predominantly on the operational side of the tourism businesses, i.e. once their construction has already been completed. Hence, they do not make explicit reference to respecting land rights prior to and during the implementation of the business. There is a need to hold hotel operators accountable for land grabs committed prior to the establishment of tourism premises and for taking advantage of an unjust legal environment.

Scholars have pointed to various other flaws in industry self-regulation, including the lack of measurable criteria (Mason and Mowforth, 1996; Scheyvens, 2002), the fragmentation of the tourism sector (Mowforth, Charlton and Munt, 2008), and the tendency of corporate actors to shift responsibility for ethical and just tourism to the tourists and/or the host governments (Scheyvens, 2002; Mowforth, Charlton and Munt, 2008). Some scholars have identified how codes of conduct act as clever marketing strategies and public relations exercises (e.g. Mason, 2016; Mowforth and Munt, 2016). Klein (2001, p. 434) asserts that the "proliferation of voluntary codes of conduct and ethical business initiatives is a haphazard and piecemeal mess of crisis management" (quoted in Mowforth and Munt, 2016, p. 212). Civil society groups, government actors and international advocacy organisations need to make clear to the industry that past and current human rights abuses (i.e. forced displacement, land dispossession, infringement on local water supplies) by tourism operators cannot be offset by corporate social responsibility practices or goodwill measures in other areas of their business operations. Tourism businesses need to be held accountable to those groups whose access to land and other natural resources and whose rights to property and housing may be directly or indirectly impacted by the instalment of their business and/or their long-term operations. National and international tourism associations should strongly encourage their members to align their operations with the UN Guiding Principles on Business and Human Rights and should expel those members that have been found responsible for human rights abuses (e.g. land grabs, evictions).

Environmental and social safeguards of international financial institutions and their relevance for tourism-induced displacement and resettlement

Several international financial institutions (IFIs), such as the World Bank and the Asian Development Bank, have recognised the drastic impact that evictions and displacements induced by infrastructure and other development projects, including tourism, can have on people. The World Bank was the first multilateral development agency that adopted guidelines on involuntary resettlement more than 35 years ago. It has since made some important contributions to the overall understanding of the multiple risks associated with evictions and proposed a set of mitigation measures – so-called environmental and social safeguards – to minimise harm. These safeguards have been formally adopted by a substantial number of other IFIs, private financial institutions and corporations, yet this has not reduced the high number of forced evictions and involuntary displacements that occur each year. A major reason for this is that foreign direct investments (FDIs), including in land and other natural resources for tourism development, are still seen as a major growth strategy and a prerequisite for poverty alleviation in less developed countries. IFIs and commercial banks have repeatedly called for a further liberalisation of land markets to allow foreign ownership, which aggravates the problem of tourism-related land and resource grabs. In addition, IFIs use an increasing number of complex financing modalities, such as multi-purpose financing facilities, financial intermediary projects and finance projects with multiple subcomponents, whose displacement impacts emerge only after the project has already been approved (Price, 2015).

Another problem is that non-traditional donors and development banks from emerging economies, such as China, Brazil, India, and South Africa, have created new modalities of operation and different transparency and safeguards standards which make development-induced displacement and resettlement even more ambiguous (Price, 2015). For instance, the China-led Asian Infrastructure Investment Bank (AIIB), which began operations in January 2016, is poised to have an enormous impact on Southeast Asian countries by funding major infrastructure projects under China's ambitious Belt and Road Initiative. The AIIB has developed its own social and environmental safeguards that, among other principles, require its borrowers to only conduct 'free, prior, and informed *consultation*' (FPICon; emphasis added) with Indigenous peoples facing loss of customary ownership of land and other natural resources and/or relocation rather than to obtain 'free, prior, and informed *consent*' (FPIC;

emphasis added) from them. This appears to be in gross disregard of the UN Declaration on the Rights of Indigenous Peoples and falls short of existing international standards as enshrined, for instance, in the Equator Principles approved by international financial institutions in 2013 (see section 10.7).

To enhance environmental and social safeguards, IFIs should refrain from financing any tourism business or infrastructure project that leads to forced displacement and dispossession of legitimate landowners. They should further assume the responsibility to monitor adherence of their clients to international human rights standards, to offer legal guidance and to withdraw funding when clients fail to respect customary or codified land and resource rights in the implementation of their projects.

Equator Principles: holding the corporate financial sector accountable to tourism-related land rights infringements?

The Equator Principles (EPs) provide a comprehensive set of guidelines for both borrowers (i.e. project developers) and lenders (i.e. financial institutions) on how to assess the environmental and social impacts of projects and incorporate safeguards into project and loan agreements. They were established in 2003, have since been revised three times and have been valid in their current form of EP IV since June 2020. As of September 2020, 110 private and public financial institutions in 38 countries have voluntarily adopted the EPs and committed themselves to incorporate them into their own policies (www.equa tor-principles.com). The EPs are modelled after the International Finance Corporation's (IFC's) Performance Standards on Environmental and Social Sustainability, which came into effect in 2012 (cf. Box 11.5). The IFC is the private lending arm of the World Bank and its relatively high environmental and social risk management standards have been globally recognised. The supporters of the EPs maintain that the principles are a rare example of successful financial market self-regulation and represent substantial progress in raising financial institutions' performance in addressing social and environmental risks of development projects, including tourism. Yet it has been argued that the EPs rely solely on voluntary reporting and therefore are not a sufficiently strong hedge against dispossession and displacement in weak regulatory environments and authoritarian political settings (Wright, 2012; Price, 2015, 2018).

> **Box 11.5 The IFC's performance standards on environmental and social sustainability on which the Equator Principles are based**
>
> - Assessment and Management of Environmental and Social Risks and Impacts;
> - Labour and Working Conditions;
> - Resource Efficiency and Pollution Prevention;

- Community Health, Safety and Security;
- Land Acquisition and Involuntary Resettlement;
- Biodiversity Conservation and Sustainable Management of Living Natural Resources;
- Indigenous Peoples;
- Cultural Heritage

Source: IFC, 2012

A recent study commissioned by the United Nations Environment Programme (UNEP) on the adoption and implementation of the EPs by selected financial institutions found that the two major motives of project financiers for adopting the principles have been to (1) increase their reputation and (2) serve as guidelines for risk management strategies (Weber and Acheta, 2016). Another important finding from the study was that the EPs did not effectively change the way environmental and social issues in project finance are assessed nor did they lead to a rejection or modification of projects with regard to their environmental and social impacts. Even more concerning is the finding that in some cases financial institutions that signed up to the EPs did not even comply with their own voluntary guidelines and that most of them did not disclose their projects in reporting, thereby making it near-impossible to assess their social and environmental impacts (Weber and Acheta, 2016).

Some critics have advocated for independent external assurance of financial institutions' EP disclosures rather than just self-reporting (Macve and Chen, 2010). It has also been proposed that the EPs implement enforcement mechanisms that could include monetary fines or exclusion from the EP association, if members do not comply with some of the principles (Weber and Acheta, 2016). With regard to land acquisition and involuntary resettlement, EP member institutions could be obligated to disclose the alternatives they explored and why these were not selected. In the meantime, stakeholder pressure through media reporting, independent research and NGO advocacy appears to be the most promising way to hold financial institutions accountable for their project impacts, as they face considerable reputational risk if they do not comply with the social and environmental safeguard principles they have committed to.

Concluding remarks

This chapter has discussed the potential and shortcomings of a range of legal and regulatory instruments with relevance to tourism-related land grabbing and displacement, including international human rights law, the ILO's Indigenous and Tribal Peoples Convention, the United Nations Declaration of the Rights of Indigenous Peoples, the FAO's Voluntary Guidelines on the Responsible Governance of Tenure of Land, Fisheries and Forests, the UN Guiding Principles on Business and Human Rights, environmental and social safeguards of IFIs, and the Equator Principles. While some of these frameworks have gradually been adopted by tourism stakeholders at government, donor and corporate levels, most of them lack teeth due to their voluntary nature, the absence of measurable criteria and sanction mechanisms, and the lack of regulatory oversight at international and national level. All too often, pro-tourism rhetoric is being used to justify unethical land acquisition and displacement practices by tourism actors. The preference among corporate tourism actors for self-regulatory code of conducts – which are endorsed by international tourism lobby organisations, such as the WTTC and the UNWTO – also hinder the implementation of stricter and more enforceable legal safeguards against tourism-related land grabs and displacement.

The next and final chapter will draw the conclusions and provide a cautious outlook on the possible emergence of more just and ethical tourism practices in the Global South. It starts with a summary of the key arguments presented in this book and then discusses two contrasting future scenarios for the tourism sector in countries of the Global South against the backdrop of the ongoing COVID-19 pandemic. It will close by briefly delineating a future research agenda for tourism-related land grabbing and displacement.

References

Bauer, T. (2015) 'Human rights obligations of international hotel chains', in Gardetti, M.A. and Torres, A.L. (eds) *Sustainability in Hospitality – How Innovative Hotels are Transforming the Industry*. Routledge: London, New York, pp. 1–17.

Brüntrup, M., Scheumann, W., Berger, A., Christmann, L. and Brandi, C. (2014) 'What can be expected from international frameworks to regulate large-scale land and water acquisitions in sub-Saharan Africa?', *Law and Development Review*, 7 (2): 433–471.

Dhanarajan, S. (2015) 'Transnational state responsibility for human rights violations resulting from global land grabs' in Carter, C. and Harding, A. (eds) *Land Grabs in Asia: What Role for the Law?* Routledge: London, New York, pp. 167–186.

FAO (2012) *Voluntary Guidelines for Responsible Governance of Tenure of Land, Forestry and Fisheries*. Food and Agricultural Organization of the United Nations (FAO): Rome.

FAO (2015) *Safeguarding Land Tenure Rights in the Context of Agricultural Investment. Governance of Tenure – Technical Guide 4*. Food and Agricultural Organization of the United Nations (FAO): Rome.

Hall, R. and Scoones, I. with Henley, G. (2016) *Strengthening Land Governance: Lessons from Implementing the Voluntary Guidelines. State of the Debate Report 2016*. UKAID: London.

IFC (2012) *IFC Performance Standards on Environmental and Social Sustainability*. Available at: www.ifc.org/wps/wcm/connect/c8f524004a73daeca09afdf998895a12/IFC_Perfo rmance_Standards.pdf?MOD=AJPERES (accessed 19 January 2018).

IHRB and Tourism Concern (2012) *Frameworks for Change: The Tourism Industry and Human Rights*. Available at: www.ihrb.org/pdf/2012-05-29-Frameworks-for-Cha nge-Tourism-and-Human-Rights-Meeting-Report.pdf (accessed 19 January 2019).

ILO (n.d.) *C169 – Indigenous and Tribal Peoples Convention, 1989* (No. 169). Available at: www.ilo.org/dyn/normlex/en/f?p=NORMLEXPUB:12100:0::NO::P12100_IL O_CODE:C169 (accessed 16 January 2019).

Klein, N. (2001) *No Logo*. Flamingo: London.

Macve, R. and Chen, X. (2010) 'The "equator principles": a success for voluntary codes?', *Accounting, Auditing & Accountability Journal*, 23 (7): 890–919.

Mason, P. (2016) *Tourism Impacts, Planning and Management* (3rd edition). Routledge: London, New York.

Mason, P. and Mowforth, M. (1996) 'Codes of conduct in tourism', *Progress in Tourism and Hospitality Research*, 2: 151–167.

Mowforth, M. and Munt, I. (2016) *Tourism and Sustainability: Development, Globalisation and New Tourism in the Third World* (4th edition). Routledge: London, New York.

Mowforth, M., Charlton, C. and Munt, I. (2008) *Tourism and Responsibility: Perspectives from Latin America and the Caribbean*. Routledge: London, New York.

Neef, A. (2016) *Land Rights Matter! Anchors to Reduce Land Grabbing, Dispossession and Displacement: A Comparative Study of Land Rights Systems in Southeast Asia and the Potential of National and International Legal Frameworks and Guidelines*. Bread for the World: Berlin.

Neef, A. (2019) '*Can national and international legal frameworks mitigate land grabbing and dispossession in South-East Asia?*', in S. Price, S. and Singer, J. (eds) *Country Frameworks for Development Displacement and Resettlement Reducing Risk, Building Resilience*. Routledge: London, New York , pp. 52–70.

OHCHR (1997) *General Comment No. 7: The Right to Adequate Housing (Art. 11 (1) of the Covenant): Forced Evictions*. Available at: https://tbinternet.ohchr.org/_layouts/treatybodyexternal/Download.aspx?symbolno=INT/CESCR/GEC/6430&Lang=en (accessed 24 January 2019).

OHCHR (2011) *United Nations Human Rights Office of the High Commissioner. Guiding Principles on Business and Human Rights. Implementing the United Nations "Protect, Respect and Remedy"*. Available at: http://ww.ohchr.org/Documents/Publications/GuidingPrinciplesBusinessHR_EN.pdf (accessed 12 December 2018).

Paoloni, L. and Onorati, A. (2014) 'Regulations of large-scale acquisitions of land: The case of the Voluntary Guidelines on the Responsible Governance of Land, Fisheries and Forests', *Law and Development Review*, 7 (2): 369–400.

PATA and WINTA (2015) *Indigenous Tourism & Human Rights in Asia & the Pacific Region: Review, Analysis, & Checklists*. Available at: www.ecotourism.org.au/assets/Resources-Hub-Indigenous-Tourism/International-Indigenous-Tourism-Human-Rights-Review-Analysis-Checklists.pdf (accessed 21 December 2018).

Price, S. (2015) 'Is there a global safeguard for development displacement?' in Satiroglu, I. and Choi, N. (eds) *Development Induced Displacement and Resettlement: New Perspectives on Persisting Problems*. Routledge, London, New York, pp. 127–141.

Price, S. (2018) 'Legislative paradigm shifts for involuntary people movement: An update' in *Paradigm_Shift: People Movement*. Australian National University College of Asia and the Pacific: Canberra, pp. 26–31.

Roundtable Human Rights in Tourism (2016) *Human Rights in Tourism – An Implementation Guideline for Tour Operators* (3rd edition). Available at: www.humanrights-in-tourism.net/implementation-guidelines (accessed 26 October 2020).

Scheyvens, R. (2002) *Tourism for Development: Empowering Communities*. Pearson Education Limited: London.

Swarbrooke, J. (1999) *Sustainable Tourism Management*. CABI International: Wallingford, New York.

Tourism Concern (2012) *Water Equity in Tourism – A Human Right, A Global Responsibility*. Available at: www.tourismconcern.org.uk/wp-content/uploads/2014/10/Water-Equity-Tourism-Report-TC.pdf (accessed 22 January 2019).

Tourism Concern (2014) *Why the Tourism Industry Needs to Take a Human Rights Approach: The Business Case.* Available at: www.tourismconcern.org.uk/wp-content/uploads/2014/10/TourismConcern_IndustryHumanRightsBriefing-FIN-4.pdf (accessed 19 January 2019).

United Nations (2008) *United Nations Declaration for the Rights of Indigenous Peoples.* Available at: www.un.org/development/desa/indigenouspeoples/wp-content/uploads/sites/19/2018/11/UNDRIP_E_web.pdf (accessed 22 February 2019).

Van Huijstee, M., Ricco, V. and Ceresna-Chaturvedi, L. (2012) *How to Use the UN Guiding Principles on Business and Human Rights in Company Research and Advocacy: A Guide for Civil Society Organisations.* SOMO with CEDHA and Cividep (India): Amsterdam.

Weber, O. and Acheta, E. (2016) *The Equator Principles: Do They Make Banks More Sustainable?* Inquiry Working Paper, 16/05. United Nations Environment Programme (UNEP). Available at: http://unepinquiry.org/wp-content/uploads/2016/02/The_Equator_Principles_Do_They_Make_Banks_More_Sustainable.pdf (accessed 19 January 2019).

Wright, C. (2012) 'Global banks, the environment, and human rights: The impact of the equator principles on lending policies and practices', *Global Environmental Politics*, 12 (1): 56–77.

12 Conclusion and outlook

Summary of the main arguments

This book has discussed 31 case studies of tourism-related land grabs and displacements in various regional contexts of the Global South. These cases – along with references to many other studies that could not be presented in more detail – have shown that land grabbing, dispossession and involuntary relocation in holiday destinations in the Global South are pervasive and often systemic phenomena, involving complex alliances of government and corporate actors. Host governments – often assisted by bilateral donors, multilateral agencies and international financial institutions – prepare the ground for domestic and foreign tourism investors through tourism zoning, large-scale infrastructure development, deregulation and generous tax incentives. National and local legal frameworks and social standards are redesigned or reinterpreted to facilitate land acquisition and involuntary resettlement of local communities. The case studies provide evidence that tourism-related land grabs and displacements happen across the political spectrum – in democratic systems, semi-authoritarian regimes and countries under authoritarian rule. Existing tenure legislation also does not seem to make a strong difference; land grabs can occur in countries with a predominantly private ownership system and formalised land title register and under more informal, customary and communal forms of land and resource tenure.

Dispossession and displacement are discursively justified by emphasising the 'public purpose' of tourism development, e.g. for economic growth and job creation, preservation of natural and cultural heritage, post-conflict rehabilitation or post-disaster recovery. However, poverty, lack of political voice and legal literacy, and harassment by tourism investors and government bodies drive local people into distress sales and the acceptance of resettlement terms that are often based on incomplete or false information. Few cases are reported where resettled communities have been able to restore their livelihoods. Nevertheless, commercial banks and international financial institutions continue to finance large-scale tourism projects and sporting mega-events that lead to involuntary resettlement of Indigenous and non-indigenous communities.

As several of the examples presented in Chapters 3–9 have shown, tourism-related land grabbing and displacement do not always occur as sudden and openly violent events, but can also take more subtle forms, such as gradual gentrification and economic and cultural displacement. Communities may be deprived of the spiritual and cultural connection to their ancestral land and lose invaluable local knowledge that has been transferred over many generations. Tourism zones may gradually deplete local water supplies and enclose coastal and forest resources, often with disproportionate impacts on women. Beyond these direct adverse impacts, tourism-related land grabs and displacement can have a range of indirect effects, such as further social marginalisation, economic exploitation, and reduced self-determination and political participation of affected communities and individuals.

As discussed in Chapter 11, international legal frameworks and guiding principles can be and are being employed by human rights advocacy groups, national courts and the international justice system in attempts to strengthen local resource rights, seek redress from dispossession and eviction and protect vulnerable communities from tourism-induced land grabs and displacement. Yet, as the chapter has also shown, there are considerable shortcomings and limitations in these frameworks and principles that have hindered transformative changes within the global tourism industry to date.

While the ongoing COVID-19 pandemic has had tremendously adverse effects on the tourism industry in terms of livelihoods lost and economic damage incurred, it may provide a unique and historic window of opportunity to 'reset' global tourism and put it on a more equitable, just and ethical pathway. Yet this will require sustained efforts at local, national and global levels to confront uneven power relations in the sector and uneven mobilities between travellers and host communities. The following subsection will explore possible transformation pathways in a post-COVID-19 world.

Transformations of the tourism sector in a post-pandemic era

There is little doubt that the COVID-19 pandemic will have a transformational impact on the global tourism industry. The short-term impacts have been enormous; job losses in the tourism sector have been predicted to amount to between 100 and 200 million in 2020, and losses to global GDP from travel and tourism have been estimated to be at least US$2.6 trillion in a best-case scenario (World Travel and Tourism Council, 2020). The airline industry alone expects revenue losses of US$419 billion in 2020, with global air traffic reduced by more than 50 per cent (IATA, 2020). Tens of thousands of small businesses in the tourism and hospitality industry may have to file for bankruptcy, with destinations in the Global South that are particularly dependent on the tourism industry being the hardest hit.

Tourism scholars have been quick to call for transformative change to the global tourism industry and the re-imagining of tourism (e.g. Ateljevic, 2020; Higgins-Desbiolles, 2020; Haywood, 2020; Everingham and Chassagne, 2020; Niewiadomski, 2020). Prior to the pandemic, critical tourism scholars have already warned against the neoliberal excesses of unfettered mass tourism and overtourism (Bianchi, 2015; Avond et al., 2019; Gibson, 2019) and called for sustainable tourism practices (Butler, 1999; Farrell and Twining-Ward, 2004; Bramwell and Lane, 2014), tourism degrowth strategies (Hall, 2009; Higgins-Desbiolles et al., 2019), the use of tourism as a positive social force (Higgins-Desbiolles, 2006), responsible tourism (Harrison and Husbands, 1996; Mowforth, Charlton and Munt, 2008; Goodwin, 2011; Sharpley and Harrison, 2019) and justice tourism (Scheyvens, 2002; Higgins-Desbiolles, 2008). While these critical voices have received little attention from national governments, domestic tourism agencies and associations, multinational corporations, global tourism lobby organisations and international financial institutions, the pandemic has given rise to even more radical calls for transformational change, such as socialising tourism for social and ecological justice (Higgins-Desbiolles, 2020), tourism equity (Benjamin et al, 2020), post-development tourism such as *Buen Vivir* – 'Good Living' (Everingham and Chassagne, 2020) and regenerative tourism (Scheller, 2020).

It is beyond the scope of this book and the expertise of the author to sketch out realistic scenarios for the general transformations that the global tourism industry will undergo in a post-pandemic era. Yet, for the issues explored in this book, two contrasting outcomes can be imagined. These are discussed in the next two sub-sections.

Scenario 1. Return to business-as-usual and a new post-pandemic normalcy

Transformation sceptics, such as Hall, Scott and Gössling (2020), refer to the impact of past pandemics on the tourism sector and contend that "the juggernaut that is international tourism will roll on. For many destinations and governments, especially those with authoritarian tendencies, the focus will be on business-as-usual" (p. 591). Ionnides and Gyimóthy (2020) also acknowledge the risk that "the sector will gradually revert to the pre-crisis unsustainable growth-oriented trajectory" (p. 626). They make reference to previous crises and catastrophes, such as the 2002–2004 SARS epidemic in Asia and the 2004 Indian Ocean Tsunami, where tourism either returned very quickly to the status quo or – worse – where powerful stakeholders in the tourism industry used the disaster as an opportunity for land grabbing (see Chapter 5 in this book).

Indeed, the massive damage that COVID-19 continues to inflict on the tourism industry in many popular destinations may provide ample opportunity for risk-taking, reckless or wealthy investors to scoop up land and property from small- and medium-sized enterprises that have gone bankrupt during the pandemic. In the shadow of the global tourism crisis, investors may also grab land from Indigenous people and other local communities in new localities while governments and international watchdogs are preoccupied with crisis management in more established tourist destinations. In previous crises, transnational tourism corporations have demonstrated a high degree of adaptability and entrepreneurialism, constantly looking for newly emerging opportunities.

Governments in tourism-dependent economies may be desperate to revive their faltering tourism sectors in a post-pandemic era and could be prone to offering even more generous incentives – including more deregulation – for international tourism players, allowing them to expand their extractive and unsustainable tourism practices at the expense of the land rights and associated human rights of local populations. Furthermore, home governments may decide to bail out their transnationally operating tourism corporations – considering them as 'too big to fail' – thereby supporting a further concentration of multinational corporate power at the expense of small- and medium-sized tourism enterprises.

A new post-pandemic 'normalcy' could include an even higher securitisation of popular destinations and tourism resorts, possibly with the pretext of providing 'safe havens' for the most affluent tourists. The Tourism Alert and Action Forum (TAAF) warned that

> tourism will be keen to get back to business as usual, grabbing onto the phrase undertourism to ramp it up again. Governments will be keen to take advantage of control and surveillance capacities that are being imposed on the excuse of the crisis and to extend these further.
>
> (cited in Higgins-Desbiolles, 2020, p. 620)

Cosmopolitan tourists may prefer to be even more segregated from host communities than before the pandemic, as they have become more risk averse and may consider locals a threat to their health and security. Parts of the tourism industry are likely to cater to new demands of customers and provide new 'elite travel' opportunities. Robinson, a subsidiary of Touristik Union International (TUI), for instance, has started to offer long-stay home office 'workations' in some of its high-end resorts to keep their business afloat.

Scenario 2. Transformation of tourism into a just and equitable social force

In a much more positive scenario, the pandemic could lead to a re-evaluation and re-orientation of tourism and radical transformative changes across the entire tourism supply chain. Governments in the Global South may have come to realise that mass tourism has made their national and local economies overly dependent on international mobilities and that other sectors, such as farming, fishing and aquaculture in coastal and rural areas, have become neglected. To counter this overdependence, they may revitalise touristic areas through better integration of various local industries. Thereby tourism could become embedded in local economies in more sustainable and equitable ways and placed into the local cultural and social context. Balinese coastal dwellers, for instance, have reportedly rediscovered seaweed farming as a strategy to diversify their income since the COVID-19 pandemic has kept foreign tourists away from their island.

Governments may decide to provide a clear regulatory framework to ensure that tourism planning is done with the active involvement of all relevant stakeholder groups, including Indigenous peoples and marginalised groups. They could engage in land restitution programmes and other forms of remedial action, such as compensating communities for their losses to predatory tourism operators and reopening areas to the public that have previously been privatised and enclosed by tourism businesses. National tourism organisations and associations could foster initiatives that promote respect for local communities' rights to land and natural resources and for each individual's right to property and housing. Civil society groups could support such transformations through promoting rights-based and participatory approaches to tourism planning and development and empower local communities by strengthening their legal literacy and advocacy skills. Conservation agencies and heritage organisations may engage in new partnerships with Indigenous communities to protect wildlife and preserve cultural heritage. They could learn from Indigenous value systems that include such principles as environmental guardianship, responsibility, belonging and care (Carr, 2020).

International tourism organisations, such as the World Tourism Organization (UNTWO) and the World Travel & Tourism Council (WTTC) – which have so far "resisted substantial change despite widespread criticism and even protests and unrests" (Higgins-Desbiolles, 2020, p. 612) – could start to rethink their focus on unfettered tourism growth and instead call on the global tourism industry to adopt more inclusive approaches to tourism development in the Global South as a crucial prerequisite to support the 17 UN Sustainable Development Goals. International financial institutions could divert their funds from large-scale tourism infrastructure projects into community-based and pro-poor tourism projects and support community land titling programmes that strengthen customary rights to land and natural resources.

In a transformative post-pandemic era, tourists would increasingly aspire to travel more ethically, opt for 'slow tourism' that includes longer stays and closer engagement with host communities, and favour small family-run or community-based tourism businesses over large-scale, enclave-style hotel and resort complexes. Concerned tourists may pressure governments and tourist businesses into establishing tourism and travel policies that respect community and individual rights. Tourists committed to just and ethical tourism would gather information from reliable sources about land rights concerns associated with particular holiday destinations or certain tourism businesses. As Scheyvens (2002, p. 120) has aptly put it, "justice tourism should be about securing the human rights of those visited, enhancing their well-being and protecting their environments on their terms, and building relationships between tourists and those visited."

In such an optimistic scenario, tourism would become a "service to the public", "be accountable to the public" and be harnessed "for the empowerment and wellbeing of local communities" (Higgins-Desbiolles, 2020, p. 618). Yet, to make such a scenario more likely to emerge, it is important to first acknowledge that pre-pandemic tourism and travel – particularly in the Global South – were inherently inequitable, destructive and violent. To be part of a truly transformative change process, the global tourism industry would need to confront its own postcolonial legacy and neoliberal practices and move far beyond its usual cosmetic changes and rhetorical commitment to human rights.

Future research agendas in the field of tourism-related land grabs and displacement

This book is arguably the first attempt to provide a global perspective on tourism, land grabs and displacement. As stated in the introductory chapter, its focus on case studies from the Global South was deliberate, yet the author is well aware that tourism-related land grabs and displacement also occur in countries of the Global North. Future contributions to the field could examine historical processes of land grabs and displacement in settler states, such as Canada, the United States, Australia and New Zealand, where past and contemporary tourism developments infringe on land claimed by Indigenous populations. The impact of new tourism models, such as Airbnb, on gentrification processes in urban tourist destinations in wealthy countries and emerging economies also deserves more attention by tourism scholars.

In a post-pandemic context, more empirical work should be conducted to examine whether the economic downturn in some touristic areas of the Global South leads to processes of land restitution whereby dispossessed customary owners are able to reclaim their land and resources from tourism businesses that have gone bankrupt. It will be important to explore whether social resistance movements against historic and contemporary land grabs have gained new momentum as a result of COVID-19 and whether local protests and transnational advocacy efforts have a higher likelihood of success in a post-pandemic era.

While several case studies presented in this book allude to the gendered and racialised dimensions of tourism-related land grabs and displacement, more systematic work is required to identify the specific impacts of these human rights abuses on different social groups, with a particular focus on gender, class and ethnicity. The gendered dimensions of resistance movements and the socio-cultural and political contexts in which specific resistance identities and assemblages of resistance emerge also warrants more research.

Finally, the author hopes that this book becomes an inspiration to future generations of tourism scholars to conduct further research into the darker sides of global tourism. Truly transformative change of the feel-good industry can only be achieved if we continue to expose the hidden spaces where tourism actors exercise various forms of power and violence at the expense of the disadvantaged and marginalised groups of society.

References

Ateljevic, I. (2020) 'Transforming the (tourism) world for good and (re)generating the potential 'new normal'', *Tourism Geographies*, 22 (3): 467–475.

Avond, G., Bacari, C. Limea, I., Seraphin, H., Gowreesunkar and Mhanna, R. (2019) 'Overtourism: A result of the Janus-faced character of the tourism industry', *Worldwide Hospitality and Tourism Themes*, 11 (5): 552–565.

Benjamin, S., Dillette, A. and Alderman, D.H. (2020) '"We can't return to normal": Committing to tourism equity in the post-pandemic age', *Tourism Geographies*, 22 (3): 476–483.

Bianchi, R.V. (2009) 'The 'critical turn' in tourism studies: A radical critique', *Tourism Geographies*, 11 (4): 484–504.

Bianchi, R.V. (2015) 'Towards a new political economy of global tourism revisited' in Sharpley, R. and Telfer, D.J. (eds) *Tourism and Development: Concepts and Issues* (2nd edition). Channel View Publications: Bristol, Buffalo, Toronto, pp. 287–331.

Bramwell, B. and Lane, B. (2014) 'The "critical turn" and its implications for sustainable tourism research', *Journal of Sustainable Tourism*, 22 (1): 1–8.

Butler, R.W. (1999) 'Sustainable tourism: A state-of-the-art review', *Tourism Geographies*, 1 (1): 7–25.

Carr, A. (2020) 'COVID-19, indigenous peoples and tourism: A view from New Zealand', *Tourism Geographies*, 22 (3): 491–502.

Everingham, P. and Chassagne, N. (2020) 'Post COVID-19 ecological and social reset: Moving away from capitalist growth models towards tourism as Buen Vivir', *Tourism Geographies*, 22 (3): 555–566.

Farrell, B.H. and Twining-Ward, L. (2004) 'Reconceptualizing tourism', *Annals of Tourism Research*, 31 (2): 274–295.

Gibson, C. (2019) 'Critical tourism studies: New directions for volatile times', *Tourism Geographies*. doi:10.1080/14616688.2019.1647453.

Goodwin, H. (2011) *Taking Responsibility for Tourism*. Goodfellow Publishers: Oxford.

Hall, C.M. (2009) 'Degrowing tourism: Decroissance, sustainable consumption and steady-state tourism', *Anatolia*, 20 (1): 46–61.

Hall, C.M., Scott, D. and Gössling, S. (2020) 'Pandemics, transformations and tourism: Be careful what you wish for', *Tourism Geographies*, 22 (3) 577–598.

Harrison, D. and Husbands, W. (1996) *Practising Responsible Tourism: International Case Studies in Tourism Planning, Policy and Development*. John Wiley & Sons: Chichester.

Haywood, K.M. (2020) 'A post COVID-19 future: Tourism re-imagined and re-enabled', *Tourism Geographies*, 22 (3): 599–609.

Higgins-Desbiolles, F. (2006) 'More than an "industry": The forgotten power of tourism as a social force', *Tourism Management*, 27 (6): 1192–1208.

Higgins-Desbiolles, F. (2008) 'Justice tourism and alternative globalisation', *Journal of Sustainable Tourism*, 16 (3): 345–364.

Higgins-Desbiolles, F. (2020) 'Socialising tourism for social and ecological justice after COVID-19', *Tourism Geographies*, 22 (3): 610–623.

Higgins-Desbiolles, F., Carnicelli, S., Krolikowski, C., Wijesinghe, G. and Boluk, K. (2019) 'Degrowing tourism: Rethinking tourism', *Journal of Sustainable Tourism*, 27 (12): 1926–1944.

Ioannides, D. and Gyimóthy, S. (2020) 'The COVID-19 crisis as an opportunity for escaping the unsustainable global tourism path', *Tourism Geographies*, 22 (3): 624–632.

Mowforth, M. and Munt, I. (2016) *Tourism and Sustainability: Development, Globalisation and New Tourism in the Third World* (4th edition). Routledge: London, New York.

Mowforth, M., Charlton, C. and Munt, I. (2008) *Tourism and Responsibility: Perspectives from Latin America and the Caribbean*. Routledge: London, New York.

Niewiadomski, P. (2020) 'COVID-19: from temporary deglobalisation to a re-discovery of tourism?', *Tourism Geographies*, 22 (3): 651–656.

Scheller, M. (2020) 'Reconstructing tourism in the Caribbean: Connecting pandemic recovery, climate resilience and sustainable tourism through mobility justice', *Journal of Sustainable Tourism*. https://doi.org/10.1080/09669582.2020.1791141

Scheyvens, R. (2002) *Tourism for Development: Empowering Communities*. Pearson Education Limited: London.

Sharpley, R. and Harrison, D. (2019) '*Introduction: Tourism and development – towards a research agenda*', in Sharpley, R. and Harrison, D. (eds) *A Research Agenda for Tourism and Development*. Edward Elgar: Cheltenham, Northampton, MA, pp. 1–34.

Index